Springer Series on
SIGNALS AND COMMUNICATION TECHNOLOGY

Signals and Communication Technology

Adaptive Nonlinear System Identification: The Volterra and Wiener Model Approaches
T. Ogunfunmi
ISBN 978-0-387-26328-1

Wireless Network Security
Y. Xiao, X. Shen, and D.Z. Du (Eds.)
ISBN 978-0-387-28040-0

Satellite Communications and Navigation Systems
E. Del Re and M. Ruggieri
ISBN: 0-387-47522-2

Wireless Ad Hoc and Sensor Networks
A Cross-Layer Design Perspective
R. Jurdak
ISBN 0-387-39022-7

Cryptographic Algorithms on Reconfigurable Hardware
F. Rodriguez-Henriquez, N.A. Saqib, A. Díaz Pérez, and C.K. Koc
ISBN 0-387-33956-6

Multimedia Database Retrieval
A Human-Centered Approach
P. Muneesawang and L. Guan
ISBN 0-387-25627-X

Broadband Fixed Wireless Access
A System Perspective
M. Engels and F. Petre
ISBN 0-387-33956-6

Distributed Cooperative Laboratories
Networking, Instrumentation, and Measurements
F. Davoli, S. Palazzo and S. Zappatore (Eds.)
ISBN 0-387-29811-8

The Variational Bayes Method in Signal Processing
V. Šmídl and A. Quinn
ISBN 3-540-28819-8

Topics in Acoustic Echo and Noise Control
Selected Methods for the Cancellation of Acoustical Echoes, the Reduction of Background Noise, and Speech Processing
E. Hänsler and G. Schmidt (Eds.)
ISBN 3-540-33212-x

EM Modeling of Antennas and RF Components for Wireless Communication Systems
F. Gustrau, D. Manteuffel
ISBN 3-540-28614-4

Interactive Video
Methods and Applications
R. I Hammoud (Ed.)
ISBN 3-540-33214-6

ContinuousTime Signals
Y. Shmaliy
ISBN 1-4020-4817-3

Voice and Speech Quality Perception
Assessment and Evaluation
U. Jekosch
ISBN 3-540-24095-0

Advanced ManMachine Interaction
Fundamentals and Implementation
K.-F. Kraiss
ISBN 3-540-30618-8

Orthogonal Frequency Division Multiplexing for Wireless Communications
Y. (Geoffrey) Li and G.L. Stüber (Eds.)
ISBN 0-387-29095-8

Circuits and Systems
Based on Delta Modulation
Linear, Nonlinear and Mixed Mode Processing
D.G. Zrilic ISBN 3-540-23751-8

Functional Structures in Networks
AMLn—A Language for Model Driven Development of Telecom Systems
T. Muth ISBN 3-540-22545-5

RadioWave Propagation
for Telecommunication Applications
H. Sizun ISBN 3-540-40758-8

Electronic Noise and Interfering Signals
Principles and Applications
G. Vasilescu ISBN 3-540-40741-3

DVB
The Family of International Standards for Digital Video Broadcasting, 2nd ed.
U. Reimers ISBN 3-540-43545-X

Digital Interactive TV and Metadata
Future Broadcast Multimedia
A. Lugmayr, S. Niiranen, and S. Kalli
ISBN 3-387-20843-7

Adaptive Antenna Arrays
Trends and Applications
S. Chandran (Ed.)
ISBN 3-540-20199-8

Digital Signal Processing
with Field Programmable Gate Arrays
U. Meyer-Baese
ISBN 3-540-21119-5

(continued after index)

Tokunbo Ogunfunmi

Adaptive Nonlinear System Identification

The Volterra and Wiener Model Approaches

Springer

Tokunbo Ogunfunmi
Santa Clara University
Santa Clara, CA
USA

ISBN 978-1-4419-3883-1 e-ISBN 978-0-387-68630-1

Printed on acid-free paper.

9 8 7 6 5 4 3 2 1

springer.com

To my parents, Solomon and Victoria Ogunfunmi

PREFACE

The study of nonlinear systems has not been part of many engineering curricula for some time. This is partly because nonlinear systems have been perceived (rightly or wrongly) as difficult. A good reason for this was that there were not many good analytical tools like the ones that have been developed for linear, time-invariant systems over the years. Linear systems are well understood and can be easily analyzed.

Many naturally-occurring processes are nonlinear to begin with. Recently analytical tools have been developed that help to give some understanding and design methodologies for nonlinear systems. Examples are references (Rugh WJ 2002), (Schetzen 1980). Also, the availability and power of computational resources have multiplied with the advent of large-scale integrated circuit technologies for digital signal processors. As a result of these factors, nonlinear systems have found wide applications in several areas (Mathews 1991), (Mathews 2000).

In the special issue of the IEEE *Signal Processing Magazine* of May 1998 (Hush 1998), guest editor Don Hush asks some interesting questions:

"....Where do signals come from?.... Where do stochastic signals come from?...." His suggested answers are, *"In practice these signals are synthesized by (simple) nonlinear systems"* and *"One possible explanation is that they are actually produced by deterministic systems that are capable of unpredictable (stochastic-like) behavior because they are nonlinear."*

The questions posed and his suggested answers are thought-provoking and lend credence to the importance of our understanding of nonlinear signal processing methods. At the end of his piece, he further writes:

Of course, the switch from linear to nonlinear means that we must change the way we think about certain fundamentals. There is no universal set of eigenfunctions for nonlinear systems, and so there is no equivalent of the frequency domain. Therefore, most of the analysis is performed in the time domain. This is not a conceptual barrier, however, given our familiarity with state-space analysis. Nonlinear systems exhibit new and different types of behavior that must be explained and understood (e.g., attractor dynamics, chaos, etc.). Tools used to differentiate such behaviors include different types of stability (e.g., Lyapunov, input/output), Lyapanov exponents (which generalize the notion of eigenvalues for a system), and the nature of the manifolds on which the state-space trajectory lies (e.g., some have fractional dimensions). Issues surrounding the development of a model are also different. For example, the choice of the sampling interval for discrete-time nonlinear systems is not governed by the sampling theorem. In addition, there is no canonical form for the nonlinear mapping that must be performed by these models, so it is often necessary to consider several alternatives. These might include polynomials, splines, and various neural network models (e.g., multilayer perceptrons and radial basis functions).

In this book, we present simple, concise, easy-to-understand methods for identifying nonlinear systems using adaptive filter algorithms well known for linear systems identification. We focus on the Volterra and Wiener models for nonlinear systems, but there are other nonlinear models as well.

Our focus here is on one-dimensional signal processing. However, much of the material presented here can be extended to two- or multi-dimensional signal processing as well.

This book is not exhaustive of all the methods of nonlinear adaptive system identification. It is another contribution to the current literature on the subject.

The book will be useful for graduate students, engineers, and researchers in the area of nonlinear systems and adaptive signal processing.

It is written so that a senior-level undergraduate or first-year graduate student can read it and understand. The prerequisites are calculus and some linear systems theory. The required knowledge of linear systems is breifly reviewed in the first chapter.

The book is organized as follows. There are three parts. Part 1 consists of chapters 1 through 5. These contain some useful background material. Part 2 describes the different gradient-type algorithms and consists of chapters 6 through 9. Part 3, which consists only of chapter 10, describes the recursive least-squares-type algorithms. Chapter 11 has the conclusions.

Chapter 1 introduces the definition of nonlinear systems. Chapter 2 introduces polynomial modeling for nonlinear systems. In chapter 3, we introduce both Volterra and Wiener models for nonlinear systems. Chapter 4 reviews the methods used for system identification of nonlinear systems. In chapter 5, we review the basic concepts of adaptive filter algorithms.

We present stochastic gradient-type adaptive system identification methods based on the Volterra model in chapter 6. In chapters 7 and 8, we present the algorithms for nonlinear adaptive system identification of second- and third-order Wiener models respectively. Chapter 9 extends this to other related stochastic-gradient-type adaptive algorithms. In chapter 10, we describe recursive-least-squares-type algorithms for the Wiener model of nonlinear system identification. Chapter 11 contains the summary and conclusions.

In earlier parts of the book, we consider only *continuous-time* systems, but similar results exist for *discrete-time* systems as well. In later parts, we consider *discrete-time* systems, but most of the results derive from *continuous-time* systems.

The material presented here highlights some of the recent contributions to the field. We hope it will help educate newcomers to the field (for example, senior undergraduates and graduate students) and also help elucidate for practicing engineers and researchers the important principles of nonlinear adaptive system identification .

Any questions or comments about the book can be sent to the author by email at togunfunmi@scu.edu or togunfunmi@yahoo.com.

Santa Clara, California Tokunbo Ogunfunmi
September 2006

ACKNOWLEDGEMENTS

I would like to thank some of my former graduate students. They include Dr. Shue-Lee Chang, Dr. Wanda Zhao, Dr. Hamadi Jamali, and Ms. Cindy (Xiasong) Wang. Also thanks to my current graduate students, including Ifiok Umoh, Uju Ndili, Wally Kozacky, Thomas Paul and Manas Deb. Thanks also to the many graduate students in the Electrical Engineering Department at Santa Clara University who have taken my graduate level adaptive signal processing classes. They have collectively taught me a lot. In particular, I would like to thank Dr. Shue-Lee Chang, with whom I have worked on the topic of this book and who contributed to some of the results reported here. Thanks also to Francis Ryan who implemented some of the algorithms discussed in this book on DSP processors.

Many thanks to the chair of the Department of Electrical Engineering at SCU, Professor Samiha Mourad, for her encouragement in getting this book published. I would also like to thank Ms. Katelyn Stanne, Editorial Assistant, and Mr. Alex Greene, Editorial Director at Springer for their valuable support.

Finally, I would like to thank my family, Teleola, Tofunmi, and Tomisin for their love and support.

CONTENTS

Chapter 1

INTRODUCTION TO NONLINEAR SYSTEMS

Why Study Nonlinear Systems?

Introduction

The subject of this book covers three different specific academic areas: nonlinear systems, adaptive filtering and system identification. In this chapter, we plan to briefly introduce the reader to the area of nonlinear systems.

The topic of system identification methods is discussed in chapter 4. The topic of adaptive (filtering) signal processing is introduced in chapter 5.

Before discussing nonlinear systems, we must first define a linear system, because any system that is *not* linear is obviously *nonlinear*.

Most of the common design and analysis tools and results are available for a class of systems that are called linear, time-invariant (LTI) systems. These systems are obviously well studied (Lathi 2000), (Lathi 2004), (Kailath 1979), (Philips 2003), (Ogunfunmi 2006).

1.1 Linear Systems

Linearity Property

Linear systems are systems whose outputs depend *linearly* on their inputs. The system property of linearity is based on two principles: (1) superposition and (2) homogeneity.

In the following, we shall consider only *continuous-time* systems, but similar results exist for *discrete-time* systems as well. These follow as simple extensions.

A system obeys the principle of superposition if the outputs from different inputs are additive: for example, if the output $y_1(t)$ corresponds to input $x_1(t)$ and the output $y_2(t)$ corresponds to input $x_2(t)$. Now if the system is subjected to an additive input $x(t) = (x_1(t) + x_2(t))$ and the corresponding output is $y(t) = (y_1(t) + y_2(t))$, then the system obeys the superposition principle. See figure 1-1.

A system obeys the principle of homogeneity if the output corresponding to a scaled version of an input is also scaled by the same scaling factor: for example, if the output y(t) corresponds to input x(t). Now if we apply an input ax(t) and we get an output ay(t), then the system obeys the principle of homogeneity. See figure 1-2.

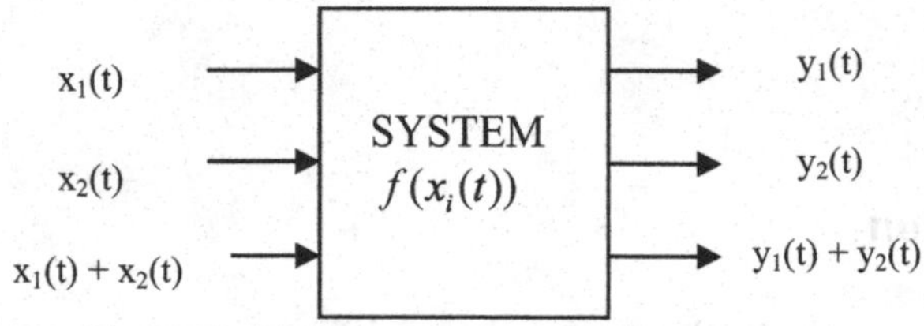

Figure 1-1. Superposition principle of linear systems

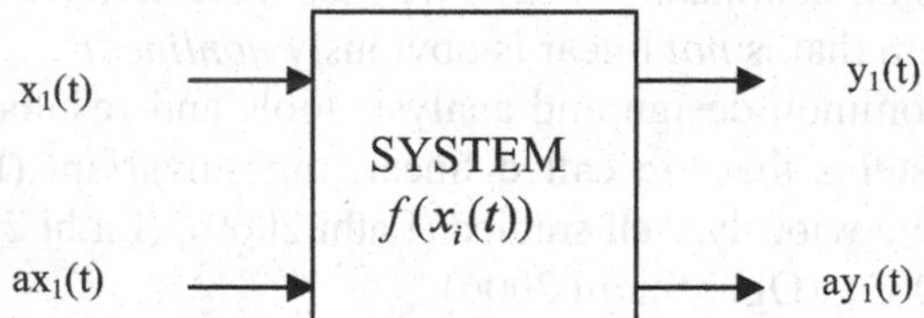

Figure 1-2. Homogeneity principle of linear systems

Both of these principles are essential for a linear system. This means if we apply the input $x(t) = (ax_1(t) + bx_2(t))$, then a linear system will produce the corresponding output $y(t) = (ay_1(t) + by_2(t))$.

In general, this means if we apply sums of scaled input signals (for example, $x(t) = ax1(t) + bx2(t) + cx3(t)$) to a linear system, then the outputs will be sums of scaled output signals $y(t) = (ay1(t) + by2(t) + cy3(t))$ where each part of the output sum corresponds respectively to each part of the input sum. This applies to any finite or infinite sum of scaled inputs. This means:

If

$$x(t) = \sum_{i=1}^{P} a_i x_i(t),$$

then

$$y(t) = \sum_{i=1}^{P} a_i y_i(t),$$

where

$$y_i(t) = f(x_i(t))$$

Examples:

Determine if the following systems are linear/nonlinear y(t) = f(x(t)):

1. $y(t) = f(x) = x^2(t)$ (nonlinear)
2. $y(t) = f(x) = x(t) + a$ (nonlinear)
3. $y(t) = f(x) = t\,x(2t)$ (linear)
4. $y(t) = f(x) = x(t),\ t < 0$
   $= -x(t),\ t >= 0$ (linear)
5. $y(t) = f(x) = |\,x(t)\,|$ (nonlinear)
6. $y(t) = f(x) = \int_{-\infty}^{t} x(\tau)d\tau$ (linear)

Time-invariant Property

A system S: S[x(t)] is time-invariant (sometimes called stationary) if the characteristics or properties of the system do not vary or change with time.

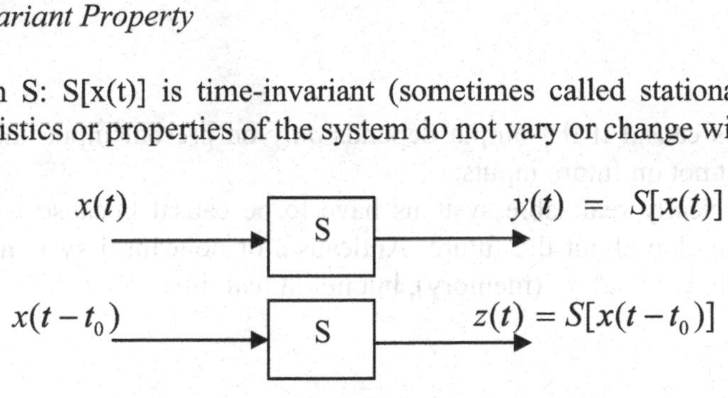

Figure 1-3. Time-invariant property of systems

If after shifting y(t), the result $y(t-t_0)$ equals $S[x(t-t_0)] = z(t)$, then the system is time-invariant. See figure 1-3.

Examples:

Determine if the following systems are time-varying/time-invariant y(t) = f(x(t)):

1. $y(t) = f(x) = 2\cos(t)\, x^2(t-4)$ (time-varying)
2. $y(t) = f(x) = 3x(4t-1)$ (time-varying)
3. $y(t) = f(x) = 5x^2(t)$ (time-invariant)
4. $y(t) = f(x) = t\, x(2t)$ (time-varying)

We will not concern ourselves much with the time-invariance property in this book. However, a large class of systems are the so-called linear, time-invariant (LTI) systems. These systems are well studied and have very interesting and useful properties like convolution, impulse response, etc.

Properties of LTI Systems

Most of the properties of linear, time-invariant (LTI) systems are due to the fact that we can represent the system by differential (or difference) equations. Such properties include: impulse response, convolution, duality, stability, scaling, etc.

The properties of linear, time-invariant system do not in general apply to nonlinear systems. Therefore we cannot necessarily have characteristics like impulse response, convolution, etc. Also, notions of stability and causality are defined differently for nonlinear systems.

We will recall these definitions here for *linear* systems.

Causality

A system is causal if the output depends *only* on present input and/or past outputs, but not on future inputs.

All physically realizable systems have to be causal because we do not have information about the future. Anticausal or noncausal systems can be realized only with delays (memory), but not in real time.

Stability

A system is bounded-input, bounded-output (BIBO) stable if a bounded input x(t) leads to a bounded output y(t). A BIBO stable system that is causal will have all its poles in the left hand side (LHS) of the s-plane. Similarly, a BIBO stable discrete-time system that is causal will have all its poles inside the unit-circle in the z-plane. We will not concern ourselves here with other types of stability.

Memory

A system is memoryless if its output at time t_0 depends *only* on input at time t_0. Otherwise, if the output at time t_0 depends on input at times $t < t_0$ or $t > t_0$, then the system has memory.

We focus on the subclass of memoryless nonlinear systems in developing key equations in chapter 2.

Representation Using Impulse Response

The impulse function, $\delta(t)$, is a generalized function. It is an ideal function that does not exist in practice. It has many possible definitions.

It is idealized because it has zero width and infinite amplitude. An example definition is shown in figure 1-4 below. It is the limit of the pulse function p(t) as its width Δ (delta) goes to zero and its amplitude $1/\Delta$ goes to infinity. Note that the area under p(t) is always 1 (Δ x $1/\Delta = 1$).

$$\delta(t) = \lim_{\Delta \to 0} p(t)$$

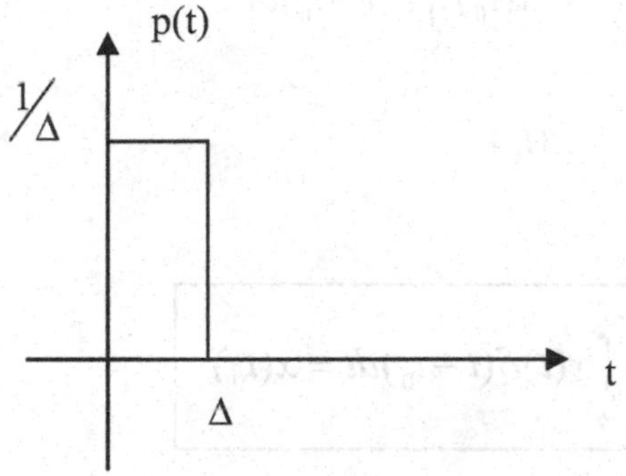

Figure 1-4. Impulse function representation

Another generalized function is the ideal rectangular function, $\Pi\left(\frac{t}{T}\right) = rect(t)$

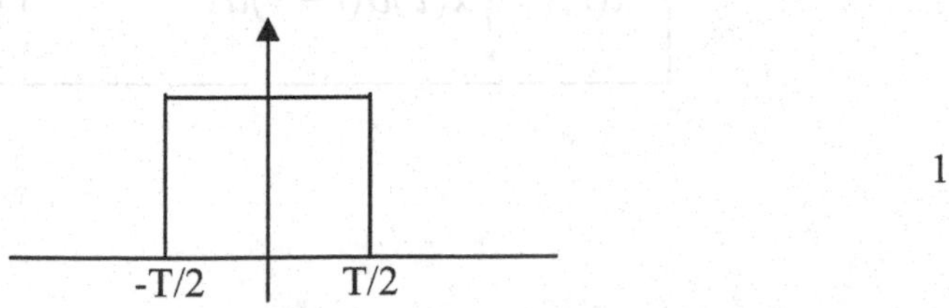

Unit Impulse:

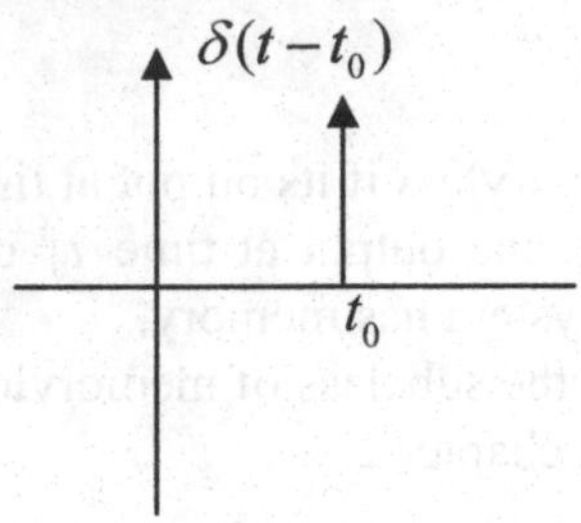

Note that $\int_{-\infty}^{\infty} \delta(\tau)d\tau = 1$ and $\int_{-\infty}^{\infty} \delta(t-t_0)dt = 1$

For all t:

$\int_{-\infty}^{t} \delta(\tau)d\tau = u(t)$ and $X(t)\delta(t) = X(0)\delta(t)$

Sifting $\quad X(t)\delta(t-t_0) = X(t_0)\delta(t-t_0)$

$$\int_{-\infty}^{\infty} x(t)\delta(t-t_0)dt = x(t_0)\int_{-\infty}^{\infty} \delta(t-t_0)dt$$

$$= x(t_0)$$

Sifting property

$$\boxed{\int_{-\infty}^{\infty} x(t)\delta(t-t_0)dt = x(t_0)}$$

It is a consequence of the sifting property that any arbitrary input signal $x(t)$ can be represented as a sum of weighted $x(\tau)$ and shifted impulses $\delta(t-\tau)$ as shown below:

$$x(t) = \int_{-\infty}^{\infty} x(\tau)\delta(t-\tau)d\tau \qquad (1.1)$$

Convolution

Recall that the impulse response is the system response to an impulse.

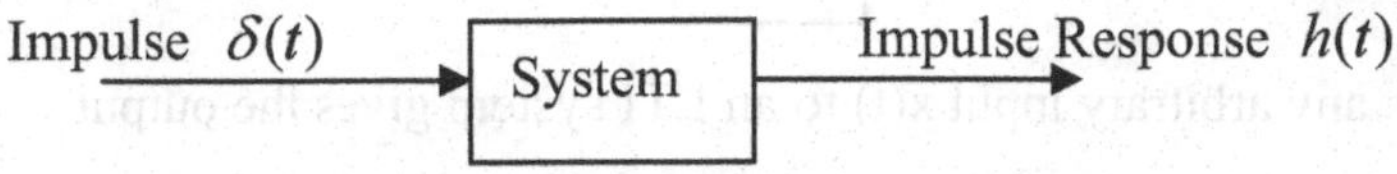

Recall the property of the impulse function that any arbitrary function x(t) can be represented as

$$x(t) = \int_{-\infty}^{\infty} x(\tau)\delta(t-\tau)d\tau$$

Any continuous-time signal can be represented by

$$x(t) = \int_{-\infty}^{\infty} x(\tau)\delta(t-\tau)d\tau$$

If the system is S, an impulse input gives the response:

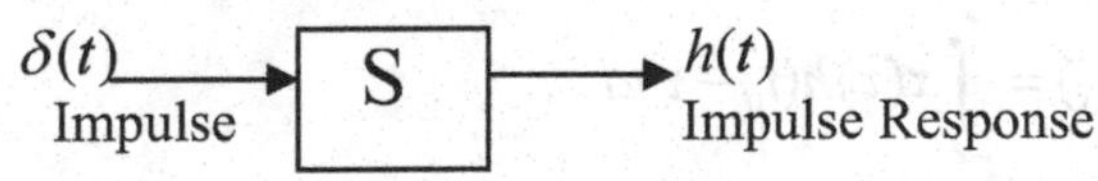

and a shifted impulse input gives the response:

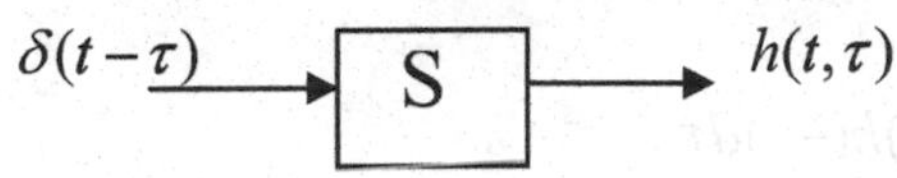

If S is linear,

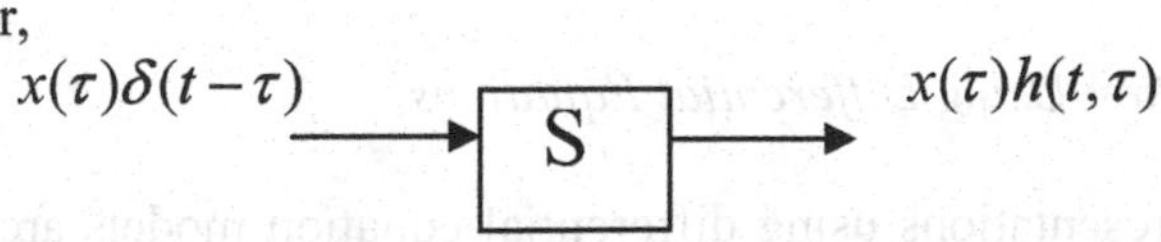

If S is time-invariant, then

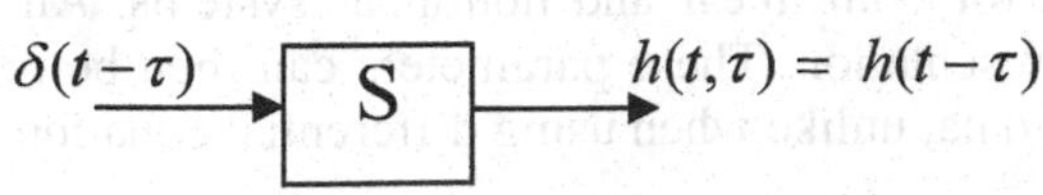

If S is linear and time-invariant, then

$$x(\tau)\delta(t-\tau) \rightarrow \text{LTI} \rightarrow x(\tau)h(t-\tau)$$

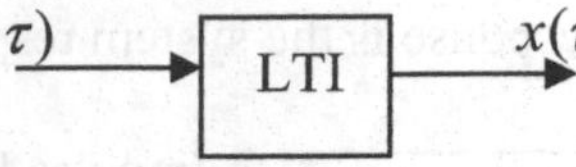

Therefore, any arbitrary input x(t) to an LTI system gives the output

$$x(t) = \int_{-\infty}^{\infty} x(\tau)\delta(t-\tau)d\tau \qquad\qquad y(t) = \int_{-\infty}^{\infty} x(\tau)h(t-\tau)d\tau \qquad (1.2)$$

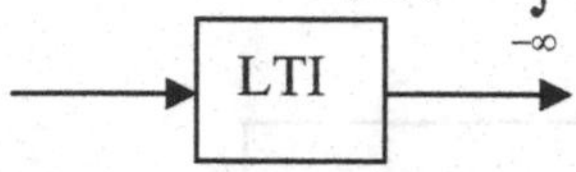

$$y(t) = x(t) * h(t)$$

$$y(t) = \int_{-\infty}^{\infty} x(\tau)h(t-\tau)d\tau$$

Note: For any time $t = t_0$, we can obtain

$$y(t_0) = \int_{-\infty}^{\infty} x(\tau)h(t_0-\tau)d\tau,$$

and for any time $t = 0$, we can obtain

$$y(0) = \int_{-\infty}^{\infty} x(\tau)h(-\tau)d\tau.$$

Representation Using Differential Equations

System representations using differential equation models are very popular. A system does not have to be represented by differential equations. There are other possible parametric and nonparametric representations or model structures for both linear and nonlinear systems. An example is the state-space representation. These parameters can then be reliably estimated from measured data, unlike when using differential equation representations.

Differential equation models: The general *n*th order differential equation is

$$a_0 y(t) + a_1 y'(t) + \ldots\ldots + a_{n-1}\frac{d^{n-1}}{dt^{n-1}}y(t) + a_n\frac{d^n}{dt^n}y(t)$$

$$= b_0 x(t) + b_1\frac{d}{dt}x(t) + \ldots\ldots + b_{m-1}\frac{d^{m-1}}{dt^{m-1}}x(t) + b_m\frac{d^m}{dt^m}x(t) \qquad m \neq n \qquad (1.3)$$

The number *n* determines the order of the differential equation. Another format of the same equation is

$$\sum_{k=0}^{n} a_k \frac{d^k}{dt^k}y(t) = \sum_{k=0}^{m} b_k \frac{d^k}{dt^k}x(t), \quad m \neq n, \qquad (1.4)$$

Consider the first order differential equation

$$\frac{d}{dt}y(t) - ay(t) = bx(t),$$

$y(t) = y_c(t) + y_p(t)$ $\qquad$ $y_c(t)$ - solution $w/\, x(t) = 0$

$y_p(t)$ - solution $w/\, x(t) \neq 0$

particular solution

Complementary/natural

$$H(s) = \frac{1}{s+a} \qquad h(t) = e^{at} \qquad \frac{d}{dt}y_c(t) = cse^{st}$$

$y_c(t)$ always has to have the form $y_c(t) = ce^{st}$ $y_c(t) = -ae^{st}$

For the general *n*th order differential equation,

$$\sum_{k=0}^{n} a_k \frac{d^k y(t)}{dt^k} = \sum_{k=0}^{m} b_k \frac{d^k x(t)}{dt^k}$$

assume $x(t) = Xe^{st}$, find y_{ss}, the steady-state response.

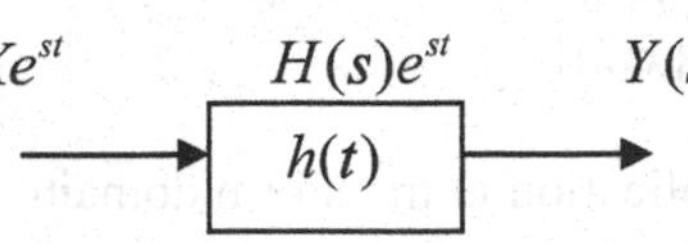

The Laplace Transform of h(t) is the transfer function.

$H(s) = L\{h(t)\}$ = Transfer function

Also, the ratio of Y(s) to X(s) is the transfer function.

$$H(s) = \frac{Y(s)}{X(s)} = \frac{b_0 + b_1 s + \ldots.. + b_m s^m}{a_0 + a_1 s + \ldots.. + a_n s^n} = \text{Transfer function} \qquad (1.5)$$

By partial fraction expansion or other methods,

$$Y(s) = H(s)X(s) = \frac{K_1}{(s-p_1)} + \frac{K_2}{(s-p_2)} + \ldots\ldots + \frac{K_n}{(s-p_n)} + \ldots\ldots + \frac{K_d}{(s-p_d)}$$

$p_1, p_2, p_3, \ldots\ldots, p_n, \ldots\ldots, p_d$, d>n, are the poles of the function Y(s).

This will give us the total solutions y(t) for any input x(t).

For x(t) = 0, $p_1, p_2, p_3, \ldots\ldots, p_n$ are the poles of transfer function H(s).

These poles are the solutions of the characteristic equation

$B(s) = b_0 + b_1 s + \ldots.. + b_m s^m = 0$.

The zeros are the solutions of the characteristic equation

$A(s) = a_0 + a_1 s + \ldots.. + a_n s^n = 0$.

$z_1, z_2, z_3, \ldots\ldots, z_n$ are the zeros of the transfer function H(s).

Representation Using Transfer Functions

Use of Laplace Transforms for continuous-time (and Z-transforms for discrete-time) LTI systems:

$$H(s) = \frac{Y(s)}{X(s)} = L\{h(t)\} = \int_{-\infty}^{\infty} h(t)e^{-st}\,dt \qquad (1.6)$$

$$Y(s) = L\{y(t)\}$$
$$X(s) = L\{x(t)\}$$

Convolution in time domain = multiplication in transform domain:

$$
\begin{aligned}
y(t) &= x(t) * h(t) \\
Y(s) &= X(s)H(s)
\end{aligned}
\tag{1.7}
$$

1.2 Nonlinear Systems

Nonlinear systems are systems whose outputs are a *nonlinear* function of their inputs.

Linear system theory is well understood and briefly introduced in section 1.1. However, many naturally-occurring systems are *non*linear. Nonlinear systems are still quite "mysterious." We hope to make them less so in this section.

There are various types of nonlinearity. Most of our focus in this book will be on nonlinear systems that can be adequately modeled by polynomials. This is because it is not difficult to extend the results (which are many) of our study of linear systems to the study of nonlinear systems.

A polynomial nonlinear system represented by the infinite Volterra series can be shown to be time-invariant for every order except for the zeroth order (Mathews 2000).

We will see later that the set of nonlinear systems that can be adequately modeled, albeit approximately, by polynomials is large. However, there are some nonlinear systems that cannot be adequately so modeled, or will require too many coefficients to model them.

Many polynomial nonlinear systems obey the principle of superposition but not that of homogeneity (or scaling). Therefore they are nonlinear. Also, sometimes linear but time-varying systems exhibit nonlinear behavior.

It is easy to show that if x(t) = c u(t), then the polynomial nonlinear system will not obey the homogeneity principle required for linearity. Therefore the output due to input x(t) = c u(t) will not be a scaled version of the output due to input x(t) = u(t). Similarly, the output may not obey the additivity or superposition property.

In addition, it is easy to show that time-varying systems lead to noncausal systems.

It is also true that for linear, time-invariant systems, the frequency components present in the output signal are the same as those present in the input signal. However for nonlinear systems, the frequencies in the output signal are not typically the same as those present in the input signal. There are signals of other new frequencies. In addition, if there is more than one input sinusoid frequency, then there will be intermodulation terms as well as the regular harmonics of the input frequencies.

Moreover, the output signal of discrete-time nonlinear systems may exhibit aliased components of the input if it is not sampled at much higher than the Nyquist rate of twice the maximum input signal frequency.

Based on the previous section, we now wish to represent the input/output description of nonlinear systems. This involves a simple generalization of the representations discussed in the previous section.

Corresponding to the real-valued function of n variables $h_n(t_1,t_2,....,t_n)$ defined for $t_i = -\infty$ to $+\infty, i = 1,2,....,n$ and such that $h_n(t_1,t_2,....,t_n) = 0$ if any $t_i < 0$, (which implies causality), consider the input-output relation

$$y(t) = \int_{-\infty}^{\infty}\int_{-\infty}^{\infty}...\int_{-\infty}^{\infty} h(\tau_1,\tau_2,....,\tau_n)x(t-\tau_1)x(t-\tau_2)....x(t-\tau_n)d\tau_1 d\tau_2....d\tau_n \tag{1.8}$$

This is the so-called *degree-n homogeneous system* (Rugh WJ 2002) and is very similar to the convolution relationship defined earlier for linear, time-invariant systems. However this system is not linear.

An infinite sum of homogeneous terms of this form is called a polynomial Volterra series. A finite sum is called a *truncated* polynomial Volterra series. Later we will present details of the nonlinear Volterra series model.

Practical Examples of Nonlinear Systems

There are many practical examples of nonlinear systems. They occur in diverse areas such as biological systems (e.g. neural networks, etc.), communication systems (e.g., channels with nonlinear amplifiers, etc.), signal processing (e.g., harmonic distortion in loudspeakers and in magnetic recording, perceptually-tuned signal processing, etc.).

This Volterra polynomial model can be applied to a variety of applications in

- system identification (Koh 1985, Mulgrew 1994, Scott 1997);
- adaptive filtering (Koh 1983, Fejzo 1995, Mathews 1996, Fejzo 1997);
- communication (Raz 1998);
- biological system modeling (Marmarelis 1978, Marmarelis 1993, Zhao 1994);
- noise canceling (Benedetto 1983);
- echo cancellation (Walach 1984);
- prediction modeling (Benedetto 1979).

The nonlinearity in the nonlinear system may take one or more of many different forms. Examples of nonlinearities are:

- smooth nonlinearities (which can be represented by polynomial models. Polynomial-based nonlinear systems include quadratic filters and bilinear filters);

- multiple-valued nonlinearities, e.g., hysteresis (as used in neural networks;
- non-smooth or nonlinearities with discontinuities;
- homomorphic systems (as used in speech processing).

To determine if our polynomial nonlinear model is adequate, we perform a nonlinearity test (Haber 1985).

It is also possible to use multi-dimensional linear systems theory to analyze Volterra series-based nonlinear systems. (For more details, see Rugh WJ 2002 and Mathews 2000.)

Many nonlinear systems arise out of a combination of a linear system and some nonlinear functions. For example, multiplicative combinations, etc. (See Rugh WJ 2002 for examples.)

Some of the other models are discussed in (Westwick K 2003). These include Volterra, Wiener, generalized Wiener, Wiener-Bose, Hammerstein (also known as NL [nonlinear-linear]), LNL (dynamic linear-static nonlinear-dynamic linear or linear-nonlinear-linear or Wiener-Hammerstein), NLN (nonlinear-linear-nonlinear cascade), parallel-cascade, etc.

In chapter 2 we will explain some of the properties of the Wiener model compared to others such as the Hammerstein model.

Nonlinear Volterra Series Expansions

Nonlinear signal processing algorithms have been growing in interest in recent years (Billigs 1980, Mathews 1991, Bershad 1999, Mathews 2000, Westwick K 2003). Numerous researchers have contributed to the development and understanding of this field. To describe a polynomial nonlinear system with memory, the Volterra series expansion has been the most popular model in use for the last thirty years. The Volterra theory was first applied by (Wiener 1942). In his paper, he analyzed the response of a series RLC circuit with nonlinear resistor to a white Gaussian signal. In modern digital signal processing fields, the *truncated* Volterra series model is widely used for nonlinear system representations. This model can be applied to a variety of applications, as seen earlier.

The use of the Volterra series is characterized by its power expansion-like analytic representation that can describe a broad class of nonlinear phenomena.

The continuous-time Volterra filter is based on the Volterra series, and its output y(n) depends linearly on the filter coefficients of zeroth-order, linear, quadratic, cubic and higher-order filter input x(n). It can be shown (Rugh WJ 2002) as:

$$y(t)=h_0+\int_{-\infty}^{\infty}h_1(\tau_1)x(t-\tau_1)d\tau_1+\int_{-\infty}^{\infty}\int_{-\infty}^{\infty}h_2(\tau_1,\tau_2)x(t-\tau_1)x(t-\tau_2)d\tau_1d\tau_2+....$$

$$+\int_{-\infty}^{\infty}\int_{-\infty}^{\infty}...\int_{-\infty}^{\infty}h_n(\tau_1,\tau_2,.....,\tau_n)x(t-\tau_1)x(t-\tau_2)....x(t-\tau_n)d\tau_1d\tau_2....d\tau_n+...... \tag{1.9}$$

where h_0 is a constant and $h_j(\tau_1,\tau_2,.....,\tau_j),\quad 1\le j\le\infty$ is the set of *j*th-order Volterra kernel coefficients defined for $\tau_i=-\infty$ to $+\infty, i=1,2,....,n,.......$ We assume $h_j(\tau_1,\tau_2,.....,\tau_j)=0$, if any $\tau_i<0,\ 1\le i\le j$ (which implies causality).

Homogeneous systems can arise in engineering applications in two possible ways, depending on the system model (Rugh WJ 2002, Mathews 1991).

The first involves physical systems that arise naturally as interconnections of linear subsystems and simple smooth nonlinearities. These can be described as *homogeneous systems*. For *interconnection structured systems* such as this, it is often easy to derive the overall system kernel from the subsystem kernels simply by tracing the input signal through the system diagram.

Homogeneous systems can also arise with a state equation description of a nonlinear system (Brogan 1991, Rugh WJ 2002). Nonlinear compartmental models of this type lead to the *bilinear state equations* such as

$$\begin{aligned}\dot{x}(t)&=Ax(t)+Dx(t)u(t)+bu(t)\\ y(t)&=cx(t),\ \ t\ge 0,\ x(0)=x_0\end{aligned} \tag{1.10}$$

where $x(t)$ is the $n \times 1$ state vector, and $u(t)$ and $y(t)$ are the scalar input and output signals respectively. Bilinear state equations are not the focus of this book, but we realize it is an alternative to polynomial models of the representation of nonlinear systems.

The causal discrete-time Volterra filter is similarly based on the Volterra series and can be shown as described by (Mathews 1991, Mathews 2000):

$$\mathrm{y(n)=h_0+\sum_{k_1=0}^{\infty}h_1(k_1)\,x(n-k_1)\ +\ \sum_{k_1=0}^{\infty}\sum_{k_2=0}^{\infty}h_2(k_1,k_2)\,x(n-k_1)\,x(n-k_2)\ +}$$

$$\mathrm{...\ +\ \sum_{k_1=0}^{\infty}...\sum_{k_p=0}^{\infty}h_p(k_1,...,k_p)\,x(n-k_1)...\,x(n-k_p)\ +\ ...} \tag{1.11}$$

where h_0 is a constant and $\{h_j(k_1, ..., k_j),\ 1\le j\le\infty\}$ is the set of jth-order Volterra kernel coefficients. Unlike the case of linear systems, it is difficult to characterize the nonlinear Volterra system by the system's unit impulse response. And as the order of the polynomial increases, the number of Volterra parameters increases rapidly, thus making the computational complexity extremely high. For simplicity, the truncated Volterra Series is most often considered in literature. The M-sample memory *p*th-order truncated Volterra Series expansion is expressed as:

$$y(n) = h_0 + \sum_{k_1=0}^{M-1} h_1(k_1)x(n-k_1) + \sum_{k_1=0}^{M-1}\sum_{k_2=0}^{M-1} h_2(k_1,k_2)x(n-k_1)x(n-k_2) + \ldots + \sum_{k_1=0}^{M-1}\ldots\sum_{k_p=0}^{M-1} h_p(k_1,\ldots,k_p)x(n-k_1)\ldots x(n-k_p) \quad (1.12)$$

There are several approaches to reducing the complexity. One approach is the basis product approximation (Wiener 1965, Metzios 1994, Newak 1996, Paniker 1996), which represents the Volterra filter kernel as a linear combination of the product of some basis vectors to attempt to reduce the implementation and estimation complexity to that of the linear problem. Another approach (Rice 1980, Chen 1989, Korenberg 1991) uses the Gram-Schmidt/modified Gram-Schmidt method and the Cholesky decomposition orthogonalization procedure to search the significant model terms to reduce the computational complexity and computer memory usage. There is also some literature involving DFT (discrete Fourier transform) frequency domain analysis (Tseng 1993, Im 1996), where overlap-save and overlap-add techniques are employed to reduce the complexity of arithmetic operations. Most of the methods mentioned above are nonadaptive and are suitable for offline environments. In recent years many adaptive nonlinear filtering algorithms have been developed that can be used in real-time applications.

Properties of Volterra Series Expansions

Volterra Series Expansions have the following properties (Mathews 2000, Rugh WJ 2002):

- *Linearity with respect to the kernel coefficients*
  This property is clearly evident from equation 1.12 and the discussion in section 2.2.2 on the implementation of Volterra filters. The output of the Volterra nonlinear system is linear with respect to the kernel coefficients. It means those coefficients are being multiplied by zero'th order bias terms, first order input samples, second order input samples, third order input samples, and so on, and then summed.
- *Symmetry of the kernels and equivalent representations*
  The permutation of the indices of a Volterra series results in symmetry of the kernels, because all permutations of any number of coefficients multiply the same combinations of input samples. This

symmetry leads to a reduction in the number of coefficients required for a Volterra series representation.

- *Multidimensional convolution property*
  The Volterra model can be written as a multidimensional convolution. For example, a *p*th order Volterra kernel can be seen as a *p*-dimensional convolution. This means the methods developed for design of multidimensional linear systems can be utilized for the design of Volterra filters.
- *Stability property*
  A *p*th order Volterra kernel is bounded-input bounded-output (BIBO) stable if

$$\sum_{k_1=0}^{M-1} \cdots \sum_{k_p=0}^{M-1} \left| h_p(k_1,\ldots,k_p) \right| < \infty .$$

  This is similar to the BIBO stability condition for linear systems. However, this condition is sufficient but not necessary for Volterra kernels. For Volterra systems with separable kernels, it is a necessary and sufficient condition.
- *Kernel complexity is very high*
  A *p*th order Volterra kernel contains N^p coefficients. Even for modest N and p, the number of kernel coefficients grows exponenttially large. Using the symmetry property, the number of independent coefficients for a *p*th order kernel can be reduced to the combination

$$N_p = \binom{N+p-1}{p}.$$

  This represents a significant reduction compared to N^p .
- *Unit impulse responses of polynomial filters not sufficient to identify all the kernel elements*
  This is perhaps the most important property. Unlike linear systems, the unit impulse response is not sufficient to represent and identify all kernel elements of a polynomial filter modeled by the Volterra series. There are other methods for determining the impulse response of a *p*th order Volterra system by finding its response to *p* distinct unit impulse functions. An example of this for a homogeneous quadratic filter (2nd order Volterra kernel) is the so called bi-impulse response.

Practical Cases Where Performance of a Linear Adaptive Filter Is Unacceptable for a Nonlinear System

The major areas where nonlinear adaptive systems are common are communications image processing and biological systems. For practical cases like these, the methods presented in this book become very useful. Here are some examples:

Communications: Bit errors in high-speed communications systems are almost entirely caused by nonlinear mechanisms. Also, satellite communication channel is typically modeled as a memoryless nonlinearity.

Image processing applications: Both edge enhancement and noise reduction are desired; but edge enhancement can be considered as a highpass filtering operation, and noise reduction is most often achieved using lowpass filtering operations.

Biological systems: They are inherently nonlinear, and modeling such systems such as the human visual system requires nonlinear models.

1.3 Summary

In this chapter we have mentioned the three different specific areas covered by this book: nonlinear systems, adaptive filtering, and system identification. This chapter has been a brief introduction to the area of nonlinear systems. In chapters 4 and 5 we cover the topics of system identification and adaptive filtering respectively.

Next, in Chapter 2, we examine in more detail the polynomial models of nonlinear systems. In chapter 3, we present details of the Volterra and Wiener nonlinear models.

Chapter 2

POLYNOMIAL MODELS OF NONLINEAR SYSTEMS

Orthogonal and Nonorthogonal Models

Introduction

In the previous chapter, we introduced and defined some terms necessary for our study of nonlinear adaptive system identification methods.

In this chapter, we focus on polynomial models of nonlinear systems. We present two types of models: orthogonal and nonorthogonal.

2.1 Nonlinear Orthogonal and Nonorthogonal Models

Signals that arise from nonlinear systems can be modeled by orthogonal or nonorthogonal models. The polynomial models that we will utilize for describing nonlinear systems are mostly orthogonal. There are some advantages to using orthogonal rather than nonorthogonal models. However we will discuss the nonorthogonal models first. This will help us better understand the importance of the orthogonality requirement for modeling nonlinear systems.

Using the Volterra series, two major models have been developed to perform nonlinear signal processing.

The first model is the nonorthogonal model and is the most commonly used. It is directly based on the Volterra series called the Volterra model. The advantage of the Volterra model is that there is little or no preprocessing needed before the adaptation. But because of the statistically nonorthogonal nature of the Volterra space spanned by the Volterra series components, it is necessary to perform the Gram-Schmidt/modified Gram-Schmidt procedure or QR decomposition method to orthogonalize the inputs. This orthogonalization procedure is crucial especially for the nonlinear LMS-type algorithms and also for the nonlinear RLS-type recursive Volterra adaptive algorithms.

The second model is the orthogonal model. In contrast to the Gram-Schmidt procedure, the idea here is to use some orthonormal bases or orthogonal polynomials to represent the Volterra series. The benefit of the orthogonal model is obvious when LMS-type adaptive algorithms are applied. The orthonormal DFT–based model will be explored in this chapter. More extensions and variations of nonlinear orthogonal Wiener models will be developed in the next few chapters.

2.2 Nonorthogonal Models

2.2.1 Nonorthogonal Polynomial Models

A polynomial nonlinear system can be modeled by the sum of increasing powers of the input signal, x(n). In general, the positive powers of x(n) are $x(n), x^2(n), x^3(n), x^4(n), x^5(n), \ldots\ldots$

Let x(n) and y(n) represent the input and output signals, respectively. For a linear causal system, the output signal y(n) can be expanded as the linear combination of M-memory input signal x(n) as

$$y(n) = c_0x(n) + c_1x(n-1) + c_2x(n-2) + c_3x(n-3) + \ldots + c_{M-1}x(n-M+1)$$

$$= \sum_{k=0}^{M-1} c_k x(n-k) \tag{2.1}$$

where c_k are the filter coefficients representing the linear causal system. If the input x(n) is white Gaussian noise, this means that the statistical properties of x(n) can be completely characterized by its mean m_x and variance σ_x^2:

$$E\{x(n)\} = m_x \tag{2.2a}$$

$$E\{(x(n)-m_x)^2\} = \sigma_x^2 \tag{2.2b}$$

Then we can say that the output y(n) is a component in an orthogonal space spanned by the orthogonal elements {x(n), x(n-1), ..., x(n-M+1)}. Taking advantage of this input orthogonal property, a lot of linear adaptive algorithms have been developed (Widrow 1985, Haykin 1996, Diniz 2002, Sayed 2003).

However, the properties mentioned above are not available when the system is nonlinear even if the input is white Gaussian noise. To see this, let us revisit the truncated *p*th order Volterra series shown in equation 1.2, where the input and output relationship is given as

$$y(n) = h_0 + \sum_{k_1=0}^{M-1} h_1(k_1)x(n-k_1) + \sum_{k_1=0}^{M-1}\sum_{k_2=0}^{M-1} h_2(k_1,k_2)x(n-k_1)x(n-k_2)$$

$$\ldots\ldots + \sum_{k_1=0}^{M-1}\ldots\sum_{k_P=0}^{M-1} h_P(k_1,\ldots,k_P)x(n-k_1)\ldots x(n-k_P) \qquad (2.3)$$

Let us assume, without loss of generality, that the kernels are symmetric, i.e., $\{h_j(k_1,\ldots,k_j), 1 \le j \le P\}$ is unchanged for any of j! permutations of indices $k_1,\ldots,k_j$ (Mathews 1991). It is easy to see that we can think of a Volterra series expansion as a Taylor series with memory. The trouble in the nonlinear filtering case is that the input components which span the space are not statistically orthogonal to each other.

For example, for a first-order nonlinear system,

$$y(n) = h_0 + \sum_{k_1=0}^{M-1} h_1(k_1)x(n-k_1).$$

For a second-order nonlinear system,

$$y(n) = h_0 + \sum_{k_1=0}^{M-1} h_1(k_1)x(n-k_1) + \sum_{k_1=0}^{M-1}\sum_{k_1=0}^{M-1} h_2(k_1,k_2)x(n-k_1)x(n-k_2).$$

And for a third-order nonlinear system,

$$y(n) = h_0 + \sum_{k_1=0}^{M-1} h_1(k_1)x(n-k_1) + \sum_{k_1=0}^{M-1}\sum_{k_2=0}^{M-1} h_2(k_1,k_2)x(n-k_1)x(n-k_2)$$

$$+\sum_{k_1=0}^{M-1}\sum_{k_2=0}^{M-1}\sum_{k_3=0}^{M-1} h_3(k_1,k_2,k_3)x(n-k_1)x(n-k_2)x(n-k_3)$$

For any nonlinear system, it may be very difficult to compute Volterra model coefficients/kernels. However, for some particular interconnections of LTI subsystems and nonlinear memory-less subsystems, it is possible.

For example, see figure 2-1 and figure 2-2 below for Wiener and Hammerstein models of nonlinear systems which are interconnections of such subsystems.

In the Weiner model, the first subsystem is an LTI system in cascade with a pure nonlinear polynomial (memory-less) subsystem.

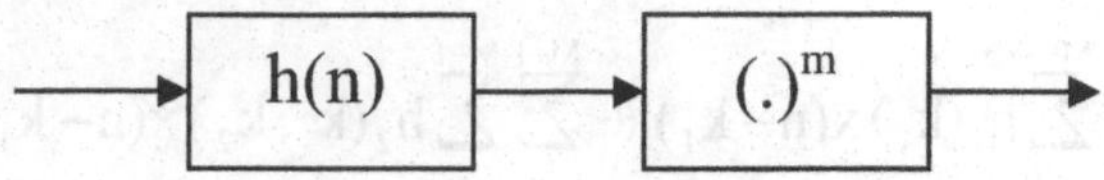

Figure 2-1. Weiner nonlinear model

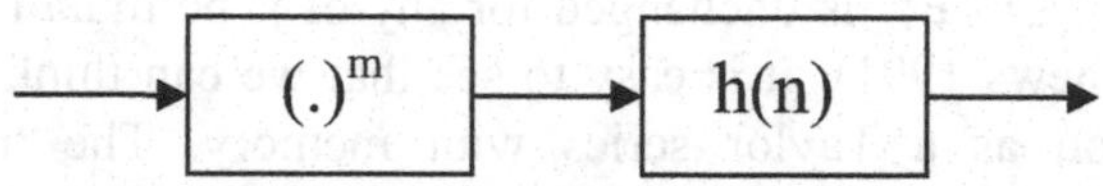

Figure 2-2. Hammerstein nonlinear model

In the Hammerstein model, the first subsystem is a pure nonlinear polynomial (memory-less) in cascade with an LTI subsystem. In the general model (figure 2-3), the first subsystem is an LTI subsystem, followed by a pure nonlinear polynomial (memory-less) then in cascade with another LTI subsystem.

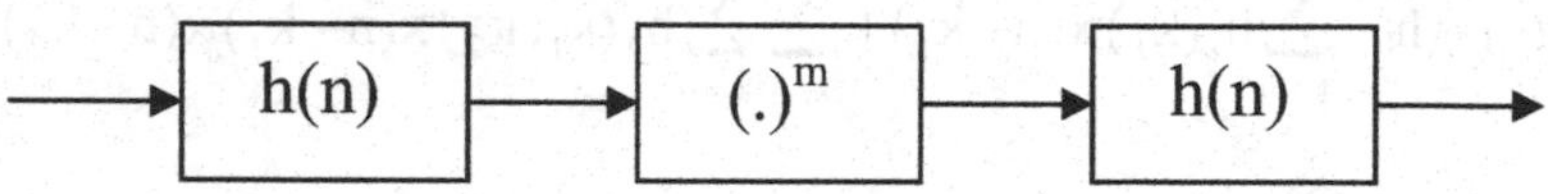

Figure 2-3. General nonlinear model

It is also possible to have other models such as a multiplicative system, which consists of LTI subsystems in parallel whose outputs are multiplied together to form the overall nonlinear system output (Rugh WJ 2002).

2.2.2 Implementation of Volterra Filters

Typically Volterra filters are implemented by interconnections of linear, time-invariant (LTI) subsystems and nonlinear, memory-less subsystems.

Volterra filters can be implemented in a manner similar to the implementation of linear filters.

For example, a zeroth-order Volterra filter is described by

$$y(n) = h_0 \tag{2.4}$$

where h_0 is the set of zeroth-order Volterra kernel coefficients. This is a trivial nonlinear system. In fact, it is a system whose output is constant irrespective of the input signal: for example, a first-order Volterra filter described by equation 2.5 where h_0 and $\{h_1(k_1)\}$ are the set of zeroth- and first-order Volterra kernel coefficients respectively.

$$y(n) = h_0 + \sum_{k_1=0}^{M-1} h_1(k_1)x(n-k_1) \tag{2.5}$$

It is easy to see that the first-order Volterra system is similar to a linear system! The difference is the zeroth-order term h_0. Without this term, equation 2.5 will be linear. Equation 2.5 can be implemented as shown in figure 2-4.

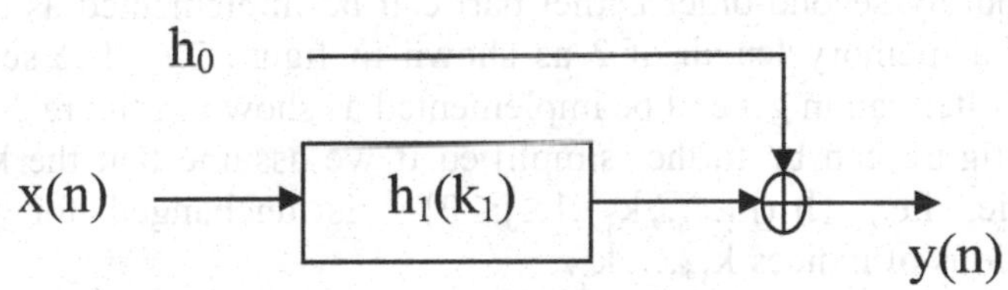

Figure 2-4. Implementation of first-order Volterra filter

For example, for a purely first-order Volterra kernel with a memory length of 2, an implementation is shown in figure 2-5.

Similarly, a second-order Volterra filter is described as follows:

$$y(n) = h_0 + \sum_{k_1=0}^{M-1} h_1(k_1)x(n-k_1) + \sum_{k_1=0}^{M-1}\sum_{k_2=0}^{M-1} h_2(k_1,k_2)x(n-k_1)x(n-k_2) \tag{2.6}$$

In equation 2.6 h_0, $\{h_1(k_1)\}$ and $\{h_j(k_1, ..., k_j), 1 \le j \le 2\}$ are the set of zeroth, first-order and second-order Volterra kernel coefficients respectively.

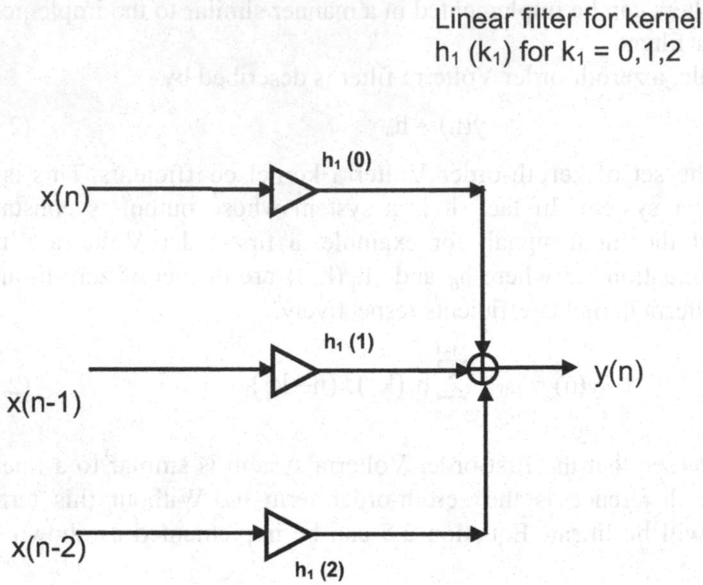

Figure 2-5. Implementation of first-order Volterra kernel ($h_1(k_1)$) for memory length of 2

The purely second-order kernel part can be implemented as a quadratic filter for a memory length of 2 as shown in figure 2-6. The second-order Volterra filter can in general be implemented as shown in figure 2-7.

This figure can be further simplified if we assume that the kernels are symmetric, i.e., $\{h_j(k_1,\ldots,k_j), 1\le j\le P\}$ is unchanged for any of j! permutations of indices $k_1,\ldots,k_j$.

A third-order Volterra filter is described by

$$y(n) = h_0 + \sum_{k_1=0}^{M-1} h_1(k_1)x(n-k_1) + \sum_{k_1=0}^{M-1}\sum_{k_2=0}^{M-1} h_2(k_1,k_2)x(n-k_1)x(n-k_2)$$
$$+\sum_{k_1=0}^{M-1}\sum_{k_2=0}^{M-1}\sum_{k_3=0}^{M-1} h_3(k_1,k_2,k_3)x(n-k_1)x(n-k_2)x(n-k_3) \tag{2.7}$$

We leave it as an exercise for the reader to determine the implementation of the purely third-order kernel of the Volterra filter.

It is easy to see from chapter 1 and from these example implementations that the Volterra filter can be implemented by interconnections of linear filter components: multipliers, adders, and delays.

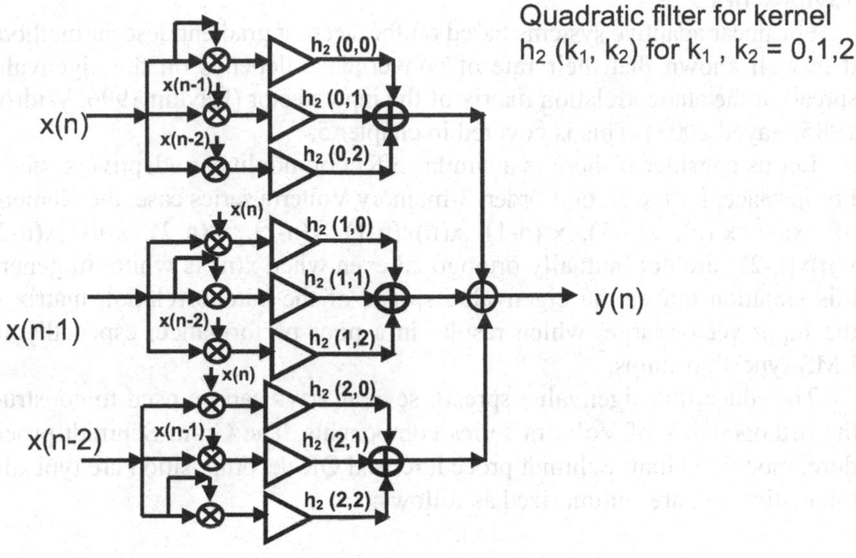

Figure 2-6. Implementation of second-order Volterra kernel ($h_2(k_1,k_2)$) for memory length of 2

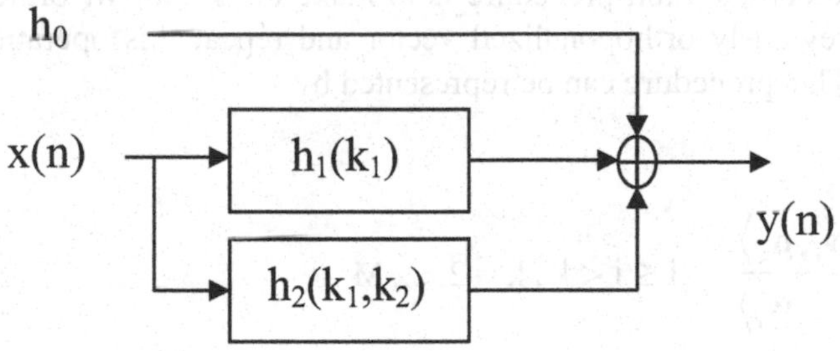

Figure 2-7. Implementation of second-order Volterra filter

For the Volterra series, assume that the kernels are symmetric, i.e.: $h_j(k_1, ..., k_j)$ is unchanged for any of j! permutations of indices $k_1, ..., k_j$. Then the number of coefficients is reduced by about half.

One can think of a Volterra series expansion as a Taylor series with memory. Recall that a Taylor series is typically defined as a representation or approximation of a function as a sum of terms computed from the evaluation of the derivatives of the function at a single point. Also recall that any smooth function without sharp discontinuities can be represented by a Taylor series.

For linear adaptive systems based on the steepest gradient descent methods, it is well known that their rate of convergence depends on the eigenvalue spread of the autocorrelation matrix of the input vector (Haykin 1996, Widrow 1985, Sayed 2003). This is covered in chapter 5.

Let us consider if there is a similar effect on nonlinear adaptive systems. For instance, for the second-order, 3-memory Volterra series case, the elements of $\{x(n), x^2(n), x(n\text{-}1), x^2(n\text{-}1), x(n)x(n\text{-}1), x(n\text{-}2), x^2(n\text{-}2), x(n\text{-}1)x(n\text{-}2), x(n)x(n\text{-}2)\}$ are not mutually orthogonal even when x(n) is white. In general this situation makes the eigenvalue spread of the autocorrelation matrix of the input vector large, which results in a poor performance, especially for LMS-type algorithms.

To reduce this eigenvalue spread, several ways can be used to construct the orthogonality of Volterra series components. The Gram-Schmidt procedure, modified Gram-Schmidt procedure, and QR decomposition are typically interesting and are summarized as follows:

2.2.3 Gram-Schmidt Orthogonalization Procedure

Assume we have a set $\{\mathbf{p}_i \mid i = 1, 2, ..., M\}$ of length m vectors and wish to obtain an equivalent orthonormal set $\{\mathbf{w}_i \mid i = 1,2, ..., M\}$ of length m vectors. The computation procedure is to make the vector $\mathbf{w}_k$ orthogonal to each k-1 previously orthogonalized vector and repeat this operation to the *M*th stage. This procedure can be represented by

$$\mathbf{w}_1 = \mathbf{p}_1$$

$$\alpha_{ik} = \frac{\langle \mathbf{w}_i, \mathbf{p}_k \rangle}{\langle \mathbf{w}_i, \mathbf{w}_i \rangle}, \quad 1 \le i < k, k = 2, ..., M \tag{2.4}$$

$$\mathbf{w}_k = \mathbf{p}_k - \sum_{i=1}^{k-1} \alpha_{ik} \mathbf{w}_i$$

where $\langle .,. \rangle$ means inner product. It is known that the Gram-Schmidt procedure is very sensitive to round off errors. In Rice (1966) it was indicated that if $\{\mathbf{p}_i \mid i = 1, 2, ...,M\}$ is ill-conditioned, using the Gram-Schmidt procedure the computed weights $\{\mathbf{w}_i \mid i = 1, 2, ..., M\}$ will soon lose their orthogonality and reorthogonalization may be needed.

2.2.4 Modified Gram-Schmidt Orthogonalization Procedure

On the other hand, the modified Gram-Schmidt procedure has superior numerical properties when operations are carried out on a computer with finite word size. The benefit is most apparent when some vectors in the set are nearly collinear. Modifying the sequence of operations in equation 2.4 slightly, the modified Gram-Schmidt procedure is to make $\mathbf{p}_{k+1}, ..., \mathbf{p}_M$ vectors orthogonal to the $\mathbf{p}_k$ vector in each stage k and repeat this operation to (M-1)th stage. The modified procedure is shown in equation 2.5 below (Brogam 1991). Initially denoting $\mathbf{p}_i^{(0)} = \mathbf{p}_i$, i = 1, ..., M, then

$$\mathbf{w}_k = \mathbf{p}_k^{(k-1)}$$

$$\hat{\mathbf{w}}_k = \frac{\mathbf{w}_k}{\langle \mathbf{w}_k, \mathbf{w}_k \rangle}, \alpha_{ki} = \langle \hat{\mathbf{w}}_k, \mathbf{p}_i^{(k-1)} \rangle \qquad \begin{array}{l} k = 1, 2, ..., M-1 \\ i = k+1, ..., M \end{array} \tag{2.5}$$

$$\mathbf{p}_i^{(k)} = \mathbf{p}_i^{(k-1)} - \alpha_{ki}\mathbf{w}_k$$

$$\mathbf{w}_M = \mathbf{p}_M^{(M-1)}$$

where $\mathbf{p}_i^{(k)}$ indicates the *i*th vector at stage k. Theoretically, identical results and the same computational complexity will be performed with both versions. The only difference is the operational sequence. However, we note that, because of the pre-processing of $\hat{\mathbf{w}}_k$, α_{ki} in equation 2.5 can be calculated with better precision than α_{ik} in equation 2.4, even if $\{\mathbf{p}_i \mid i = 1, 2, ..., M\}$ is ill-conditioned. Therefore the modified Gram-Schmidt procedure has much better numerical stability and accuracy than the Gram-Schmidt procedure.

2.2.5 QR and Inverse QR Matrix Decompositions

QR matrix decomposition is frequently used in RLS-type adaptive algorithms. This QR decomposition technique can be obtained by using the Gram-Schmidt procedure (Brogam 1991). Basically, the method is to express an $n \times m$ matrix $\mathbf{P}$ as a product of an orthogonal $n \times m$ matrix $\mathbf{Q}$ (i.e., $\mathbf{Q}^T\mathbf{Q} = \mathbf{I}$) and an upper-triangular $m \times m$ matrix $\mathbf{R}$. The Gram-Schmidt procedure is one way of determining $\mathbf{Q}$ and $\mathbf{R}$ such that $\mathbf{P} = \mathbf{QR}$. To see this, assume that the m's column vectors $\{\mathbf{p}_j \mid j= 0,1,\ldots,M-1\}$ of $\mathbf{P}$ are linearly independent. If the Gram-Schmidt procedure is applied to the set $\{\mathbf{p}_j\}$, the orthonormal set $\{\mathbf{q}_j\}$ can be obtained. The construction equations can be written as the matrix equation:

$$[\mathbf{q}_1\mathbf{q}_2 \ldots \mathbf{q}_m] = [\mathbf{p}_1\mathbf{p}_2 \ldots \mathbf{p}_m] \begin{bmatrix} \alpha_{11} & \alpha_{12} & \cdots & \alpha_{1m} \\ & \alpha_{22} & \cdots & \alpha_{2m} \\ & & \ddots & \\ & & & \alpha_{mm} \end{bmatrix} \tag{2.6}$$

where α_{ij} are the coefficients. The calculation of the α_{ij} involves inner product and norm operation as in equation 2.4. Equation 2.6 can simply be written as $\mathbf{Q} = \mathbf{PS}$. Note that $\mathbf{Q}$ need not be square, and $\mathbf{Q}^T\mathbf{Q} = \mathbf{I}$. The $\mathbf{S}$ matrix is upper-triangular and nonsingular [Brogam91]. It implies that the inverse matrix $\mathbf{S}^{-1}$ is also triangular. Therefore

$$\mathbf{P} = \mathbf{QS}^{-1} = \mathbf{QR} \tag{2.7}$$

where $\mathbf{R} = \mathbf{S}^{-1}$. Equation 2.7 is the widely used form of QR decomposition. The original column in $\mathbf{P}$ can always be augmented with additional vector $\mathbf{v}_n$ in such a way that matrix $[\mathbf{P} \mid \mathbf{V}]$ has n linearly independent columns. The Gram-Schmidt procedure can then be applied to construct a full set of orthonormal vectors $\{\mathbf{q}_j \mid j = 1,\ldots,m\}$ which can be used to find the $n \times m$ matrix $\mathbf{Q}$.

Determination of a QR decomposition is not generally straightforward. There are several computer algorithms developed for this purpose. The QR decomposition provides a good way to determine the rank of a matrix. It is also widely used in many adaptive algorithms, especially the RLS-type (Ogunfunmi 1994).

2.3 Orthogonal models

Orthogonal models are foundational and can be divided into two parts: transform-based models and orthogonal polynomial-based models. We will discuss these two parts next.

2.3.1 DFT-Based or Other Transform-Based Nonlinear Model

Wiener derived orthogonal sets of polynomials from Volterra series in Schetzen (1980, 1981), as in figure 2-8. From Wiener's theory, the $g_m[\mathbf{x}(n)]$ are chosen to be statistically orthonormal.

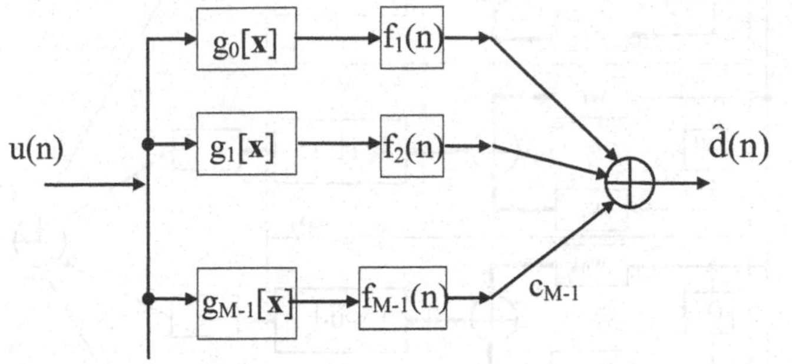

Figure 2-8. Orthonormal-based model structure

Therefore, as suggested by (Mulgrew 1994) and (Scott 1997), we can choose $g_m[\mathbf{x}(n)]$ as a set of statistical orthonormal exponential bases:

$$g_m[\mathbf{x}(n)] = \exp[j\mathbf{w}_m^T\mathbf{x}(n)] \tag{2.8}$$

For the FIR (finite impulse response) case, define the N-dimensional space vector $\mathbf{x}(n)$ as

$$\mathbf{x}(n) = [u(n), u(n-1), ..., u(n-N+1)]^T \tag{2.9}$$

where superscript $[.]^T$ indicates matrix transpose and u(n) is the input signal. For all $u(n) \in [-a, a]$, we can divide this interval into M equal divisions. Then the DFT discrete frequencies are $2\pi n/[(M+1)x_0]$, for n = 0, ..., M-1 and x_0 = 2a/M. The $\mathbf{w}_m$ is a $N \times 1$ vector whose components are chosen from the $(M+1)^N$ different combinations of discrete frequencies. For example, for a two-dimensional space the basis frequencies are chosen from two harmonically related sets. The basis function $g_m[\mathbf{x}(n)]$ can be implemented with modularity as shown in figure 2-9, where the thin solid arrows and bold arrows represent the real data and complex data flows respectively. The $g_m[\mathbf{x}(n)]$ in equation 2.8 is the statistical orthonormal basis set, which means that

$$E\{g_i^*[\mathbf{x}(n)]\,g_j[\mathbf{x}(n)]\} = \delta_{ij} \tag{2.10}$$

where δ_{ij} is the Dirac delta function.

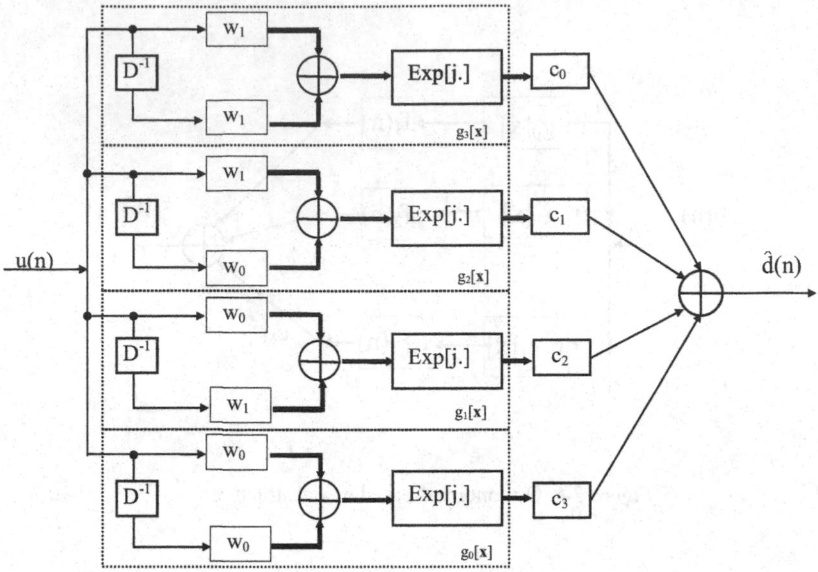

Figure 2-9. Two-dimensional nonlinear orthonormal filter modulization

To extend this to the IIR (infinite impulse response) case, the **x**(n) may include the feedback terms

$$x(n) = [\,u(n), u(n-1), \ldots, u(n-N+1), \hat{d}(n-1), \ldots, \hat{d}(n-L)\,]^T \tag{2.11}$$

Then $\mathbf{w}_m$ becomes an $(N+L)\times 1$ vector whose components are chosen from the $(M+1)^{N+L}$ different combinations of discrete frequencies. In the system identification application, the estimated $\hat{d}(n)$ can be calculated by Karhunen-Loeve-like expansion by using orthonormal functions $g_m[\mathbf{x}(n)]$ such as:

$$\hat{d}(n) = \sum_m c_m^* \, g_m[\mathbf{x}(n)] \tag{2.12}$$

where superscript $(.)^*$ means complex conjugate and c_m are the coefficients to be adapted. If we apply an adaptive algorithm, equation 2.12 can approach the desired signal d(n) in the mean square sense, which means that the mean square error $E\{|e(n)|^2\}$ is minimized:

$$\text{Min } E\{|e(n)|^2\} = \text{Min } E\{|d(n) - \hat{d}(n)|^2\} \tag{2.13}$$

The LMS algorithm is probably the most famous adaptive algorithm because of its simplicity and stable properties and is very suitable to apply to the nonlinear orthonormal filter structure. From equation 2.13 the error signals can be obtained, then the coefficients can be updated according to

$$c_m(n) = c_m(n-1) + 2g_m[\mathbf{x}(n)]e^*(n) \tag{2.14}$$

The whole architecture of nonlinear adaptive filter is shown in figure 2-10.

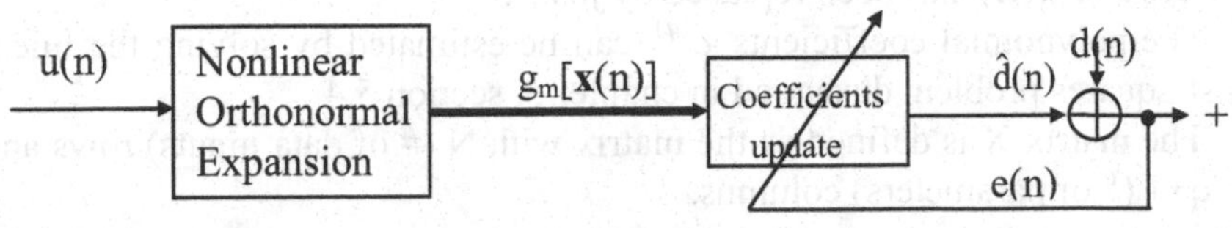

Figure 2-10. Architecture of the DFT-based nonlinear filtering

This structure is similar to (Mulgrew 1994) and (Scott 1997), but without the estimated probability density function (PDF)-divider which may cause unexpected numerical instability (Chang 1998b).

2.3.2 Orthogonal Polynomial-Based Nonlinear Models

For a memory-less nonlinear system, polynomials can be used to model the instantaneous relationship between the input and output signals. However, for a nonlinear system with memory, the Volterra series can be used to relate the instantaneous input and output signals.

In chapter 5 we will show that polynomial modeling can also be a result of solving a linear least-squares estimation problem.

The first type of polynomial model we will discuss is the power series.

Power Series

The power series (Beckmann 1973, Westwick K 2003) is a simple polynomial representation involving a sum of monomials in the input signal. The sum can be finite or infinite. The power series is typically the Taylor series of some known function.

The output y_i (or $y(n)$) is represented by a weighted sum of monomials of input signal x_i (or $x(n)$). If $y(n) = f(x(n))$, then we have an example such as:

$$f(x(n)) = \sum_{q=0}^{Q} c^{(q)} x^{q}(n)$$

$$f(x(n)) = c^{(0)} + c^{(1)}x(n) + c^{(2)}x^{2}(n) + c^{(3)}x^{3}(n) + \ldots\ldots + c^{(Q)}x^{Q}(n) \quad (2.15)$$

$$f(x) = \sum_{q=0}^{Q} c^{(q)} M^{(q)}(x) \quad (2.16)$$

where the notation $M^{(q)}(x) = x^{q}$ is introduced to represent polynomial powers and $x(n)$ has been replaced by just x.

The polynomial coefficients $c^{(q)}$ can be estimated by solving the linear least-squares problem discussed in chapter 5, section 5.4.

The matrix X is defined as the matrix with N (# of data inputs) rows and Q=q+1 (# of parameters) columns.

$$X = \begin{pmatrix} 1 & x_1 & {x_1}^2 & \cdots & {x_1}^q \\ 1 & x_2 & {x_2}^2 & \cdots & {x_2}^q \\ 1 & x_3 & {x_3}^2 & \cdots & {x_3}^q \\ \vdots & \vdots & \vdots & \ddots & \vdots \\ 1 & x_N & {x_N}^2 & \cdots & {x_N}^q \end{pmatrix} \quad (2.17)$$

This approach gives reasonable results provided there is little noise and the polynomial is of low order. However, this problem often becomes badly conditioned, resulting in unreliable coefficient estimates. This is because in power series formulations the Hessian will often have a large condition number; which means the ratio of the largest and smallest singular values will be large. As a result the estimation problem is ill-conditioned, since a small condition number of matrix X arises for two reasons:

1. The columns of X will have widely different amplitudes, particularly for high-order polynomials, unless $\sigma_x \approx 1$. As a result, the singular values of X, which are the square roots of the singular values of the Hessian, will differ widely.
2. The columns of X will not be orthogonal. This is most easily seen by examining the Hessian, $X^T X$, which will have the form

$$H = N \begin{pmatrix} 1 & E[x] & E[x^2] & \cdots & E[x^q] \\ E[x] & E[x^2] & E[x^3] & \cdots & E[x^{q+1}] \\ E[x^2] & E[x^3] & E[x^4] & \cdots & E[x^{q+2}] \\ \vdots & \vdots & \vdots & \ddots & \vdots \\ E[x^q] & E[x^{q+1}] & E[x^{q+2}] & \cdots & E[x^{2q}] \end{pmatrix} \tag{2.18}$$

Since H is not diagonal, the columns of X will not be mutually orthogonal to each other. Note that the singular values of X can be viewed as the lengths of the semi-axes of a hyper ellipsoid defined by the columns of X. Thus, nonorthogonal columns will stretch this ellipse in directions more nearly parallel to multiple columns and will shrink it in other directions, increasing the ratio of the axis lengths, and hence the condition number of the estimation problem.

Orthogonal Polynomials

We can replace the basis functions of the power series with another polynomial. The polynomial can be chosen such that it is orthonormal, and the resulting Hessian matrix is diagonal with elements of similar size.

Also the solutions of the Sturm-Liouville system of equations result in several orthogonal polynomials that are special cases. The Sturm-Liouville boundary value problem (see appendix 2A for more details) is found in the treatment of a harmonic oscillator in quantum mechanics.

Orthogonal Hermite Polynomials

Hermite polynomials result from solving the Sturm-Liouville system for a choice of parameters (see appendix 2A). The harmonic oscillator is defined by the differential equation

$$y''' - 2xy' + 2ny = 0$$

where n is a real number. For non-negative $n = 0,1,2,3,........$, the solutions of the Hermite's differential equation are referred to as Hermite polynomials, $H^n(x)$.

Hermite polynomials $H^n(x)$ can be expressed as (Efunda 2006):

$$H^n(x) = (-1)^n e^{x^2} \frac{d^n}{dx^n}(e^{-x^2}) \text{ where } n = 0,1,2,3......$$

The generating function of the Hermite polynomial is

$$e^{2tx-t^2} = \sum_{n=0}^{\infty} \frac{H^n(x)t^n}{n!}$$

It can be shown that

$$\int_{-\infty}^{\infty} e^{-x^2} H^m(x) H^n(x) dx = \begin{cases} 0, & m \neq n \\ 2^n n! \sqrt{\pi} & m = n \end{cases}$$

We note that $H_n(x)$ is even when n is even and $H_n(x)$ is odd when n is odd. Hermite polynomials form a complete orthogonal set on the interval $-\infty < x < +\infty$ with respect to the weighting function e^{-x^2}.
By using this orthogonality, a piece-wise continuous function $f(x)$ can be expressed in terms of Hermite polynomials:

$$\sum_{n=0}^{\infty} C_n H_n(x) = \begin{cases} f(x) \text{ where } f(x) \text{ is continuous} \\ \dfrac{f(x^-) + f(x^+)}{2} \text{at dis-continuous points} \end{cases}$$

where

$$C_n = \frac{1}{2^n n! \sqrt{\pi}} \int_{-\infty}^{\infty} e^{x^2} f(x) H^{(n)}(x) dx$$

This orthogonal series expansion is also known as the Fourier-Hermite series expansion or the generalized Fourier series expansion.

Orthogonal Tchebyshev Polynomials

Tchebyshev's differential equation arises as a special case in the Sturm-Liouville boundary value problem which is defined as

$$(1-x^2)y'' - xy' + n^2 y = 0$$

where n is a real number. The solutions of this differential equation are referred to as Tchebyshev's functions of degree n. For non-negative n, $n = 0,1,2,3,........$, Tchebyshev's functions are referred to as Tchebyshev's polynomials, $T^n(x)$.
Tchebyshev's polynomials $T^n(x)$ can be expressed as (Efunda 2006):

$$T^n(x) = \frac{\sqrt{1-x^2}}{(-1)^n (2n-1)(2n-3)....1} \frac{d^n}{dx^n} (1-x^2)^{n-\frac{1}{2}} \quad \text{where } n = 0,1,2,3......$$

The generating function of the Tchebyshev's polynomial is

$$\frac{1-tx}{1-2tx+t^2}=\sum_{n=0}^{\infty}T^n(x)t^n .$$

It can be shown that

$$\int_{-1}^{1}\frac{1}{\sqrt{1-x^2}}T^m(x)T^n(x)dx=\begin{cases}0, & m\neq n\\ \pi, & m=n=0\\ \frac{\pi}{2} & m=n=1,2,3.....\end{cases}$$

Note that $T_n(x)$ is even when n is even and $T_n(x)$ is odd when n is odd; and similarly for $U_n(x)$, the Tchebyshev polynomials of the second kind.

Tchebyshev polynomials form a complete orthogonal set on the interval $-1<x<+1$ with respect to the weighting function $(1-x^2)^{-1/2}$.

By using this orthogonality, a piece-wise continuous function $f(x)$ in the interval $-1<x<+1$ can be expressed in terms of Tchebyshev's polynomials:

$$\sum_{n=0}^{\infty}C_nT_n(x)=\begin{cases}f(x) \text{ where } f(x) \text{ is continuous}\\ \frac{f(x^-)+f(x^+)}{2} \text{ at dis-continuous points}\end{cases}$$

where

$$C_n=\begin{cases}\frac{1}{\pi}\int_{-1}^{1}\frac{1}{\sqrt{1-x^2}}f(x)T^{(n)}(x)dx, & n=0\\ \frac{2}{\pi}\int_{-1}^{1}\frac{1}{\sqrt{1-x^2}}f(x)T^{(n)}(x)dx, & n=1,2,3.....\end{cases}$$

This orthogonal series expansion is also known as the Fourier- Tchebyshev series expansion or a generalized Fourier series expansion.

There are other orthogonal polynomials which are not discussed here, such as Bessel, Legendre, and Laguerre polynomials (see appendix 2A for details).

2.4 Summary

In this chapter, we have focused on polynomial models of nonlinear systems. We presented the two types of models: orthogonal and nonorthogonal. Examples of orthogonal polynomials are Hermite polynomials, Legendre polynomials, and Tchebyshev polynomials.

Unlike the nonlinear nonorthogonal Volterra model, the nonlinear discrete-time Wiener model is based on an orthogonal polynomial series which is derived from the Volterra series. The particular polynomials to be used are determined by the characteristics of the input signal that we are required to model. For Gaussian, white input, the Hermite polynomials are chosen. We note that the set of Hermite polynomials is an orthogonal set in a statistical sense. This means that the Volterra series can be represented by some linear combination of Hermite polynomials. In fact every Volterra series has a unique Wiener model representation. This model gives us a good eigenvalue spread of autocorrelation matrix (which is a requirement for convergence of gradient-based adaptive filters as discussed in chapter 5), and also allows us to represent a complicated Volterra series without over-parameterization with only a few coefficients. It is interesting to note that most of the linear properties of adaptive algorithms are still preserved. By using this nonlinear model, a detailed adaptation performance analysis can be done. Further development and discussion will be presented in the next few chapters.

2.5 Appendix 2A (Sturm-Liouville System)

The general form of the Sturm-Liouville system (Beckmann 1973) is

$$\frac{\partial}{\partial x}\left[p(x)\frac{\partial y}{\partial x}\right]+[q(x)+\lambda r(x)]y=0$$

$$\begin{cases} a_1 y(a)+a_2 y^{'}(a)=0 \\ b_1 y(b)+b_2 y^{'}(b)=0 \end{cases}$$

where $a \le x \le b$.

The special cases of this system that result in orthogonal polynomials are:

Bessel Functions

For a choice of

$a=0, b=\infty, p(x)=x, q(x)=-\nu^2/x, r(x)=x$, and $\lambda=n^2$, the Sturm-Liouville equation becomes the Bessel's differential equation $\frac{\partial}{\partial x}\left[x\frac{\partial y}{\partial x}\right]+[-\frac{\nu^2}{x}+n^2 x]y=0$ which is defined in the interval $0<x<\infty$

The solutions of the Bessel's differential equation are called Bessel functions of the first kind, $J_n(x)$, which form a complete orthogonal set on the interval $0<x<\infty$, with respect to $r(x)=x$.

Legendre Polynomials

For a choice of

$a=-1, b=1, p(x)=1-x^2, q(x)=0, r(x)=1,$ and $\lambda=n(n+1),$ the Sturm-Liouville equation becomes the Legendre's differential equation $\frac{\partial}{\partial x}\left[(1-x^2)\frac{\partial y}{\partial x}\right]+n(n+1)y=0$ which is defined in the interval $-1\le x\le 1$

The solutions of the Legendre's differential equation with $n=0,1,2,......$ are called Legendre's polynomials, $P_n(x)$, which form a complete orthogonal set on the interval $-1\le x\le 1$.

Hermite Polynomials

For a choice of

$a=-\infty, b=\infty, p(x)=e^{-x^2/2}, q(x)=0, r(x)=e^{-x^2/2}$, and $\lambda=2n,$ the Sturm-Liouville equation becomes the Hermite's differential equation $\frac{\partial}{\partial x}\left[e^{-x^2/2}\frac{\partial y}{\partial x}\right]+2ne^{-x^2/2}y=0$ which is defined in the interval $-\infty\le x\le\infty$

The solutions of the Hermite's differential equation with $n=0,1,2,......$ are called Hermite's polynomials, $H_n(x)$, which form a complete orthogonal set on the interval $-\infty\le x\le\infty$ with respect to $r(x)=e^{-x^2/2}$.

Laguerre Polynomials

For a choice of

$a=0, b=\infty, p(x)=xe^{-x}, q(x)=0, r(x)=e^{-x}$, and $\lambda=n,$ the Sturm-Liouville equation becomes the Lagurre's differential equation $\frac{\partial}{\partial x}\left[xe^{-x}\frac{\partial y}{\partial x}\right]+ne^{-x}y=0$ which is defined in the interval $0<x<\infty$

The solutions of the Laguerre's differential equation with $n=0,1,2,......$ are called Laguerre's polynomials, $L_n(x)$, which form a complete orthogonal set on the interval $0<x<\infty$ with respect to $r(x)=e^{-x}$.

Tchebyshev Polynomials

For a choice of

$a=-1, b=1, p(x)=\sqrt{1-x^2}, q(x)=0, r(x)=1\Big/\sqrt{1-x^2}$, and $\lambda=n^2$, the Sturm-Liouville equation becomes the Tchebyshev's differential equation $\frac{\partial}{\partial x}\left[\sqrt{1-x^2}\frac{\partial y}{\partial x}\right]+\frac{n^2}{\sqrt{1-x^2}}y=0$ which is defined in the interval $-1<x<1$

The solutions of the Tchebyshev's differential equation are called Tchebyshev polynomilas, $T_n(x)$, which form a complete orthogonal set on the interval $-1<x<-1$, with respect to $r(x)=\frac{1}{\sqrt{1-x^2}}$.

Chapter 3

VOLTERRA AND WIENER NONLINEAR MODELS

Introduction

As shown in chapter 2, there are many approaches to nonlinear system identification. These approaches rely on a variety of different models for the nonlinear system and also depend on the type of nonlinear system.

In this book we have focused primarily on two polynomial models, the ones based on truncated Volterra series. These models are the Volterra and Wiener models. In this chapter the polynomial modeling of nonlinear systems by these two models is discussed in detail. Relationships that exist between the two models are explained and the limitations of potentially applying each model to system identification applications are explained. A historical account of the development of some of these ideas and the connection between the work of Norbert Wiener and Yuk Wing Lee is discussed in (Therrien 2002).

The success obtained in modeling a system depends upon how much information concerning the system is available. The more information that can be provided, the more accurate the system model. However, a system is often available only as a "black box" so that the relation between the input and output is difficult to determine. This is especially true for many nonlinear physical systems. The reason is that the behavior of nonlinear systems can not be characterized simply by the system's unit impulse response. Therefore, the Volterra and Wiener models are utilized to analyze such systems.

Although most previous works on the Wiener model were formulated on a Laguerre-based structure (Hashad 1994, Schetzen 1980, Fejzo 1995, Fejzo 1997), here we do not restrict ourselves to this particular constraint. With more flexible selection, a delay line version Wiener model can be developed. Consequently, some of the extensions and results can be helpful in designing a nonlinear adaptive system in the later chapters. For pure digital signal processing consideration, all the formulas derived here are based on the discrete time case under a time-invariant causal environment. Unless specifically stated, we consider the input is Gaussian white noise.

3.1 Volterra Representation

3.1.1 Zeroth- and First-Order (Linear) Volterra Model

The zeroth-order Volterra model is just a constant defined as

$$Y_0[x(n)] = h_0 \tag{3.1}$$

where $x(n)$ is the input signal and h_0 is a constant.

The first-order Volterra system is basically the same as the linear system. In other words, the linear system is a subclass of the Volterra system. Consider a general isolated linear system as shown in figure 3-1:

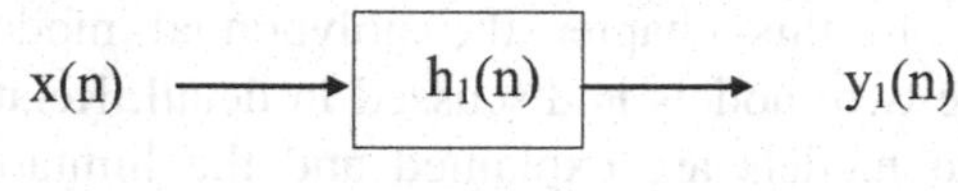

Figure 3-1. Isolated first order linear system block diagram

where the $h_1(n)$ represents the linear filter coefficients. The output $y_1(n)$ can be expressed by input $x(n)$ as:

$$y_1(n) = x(n)*h_1(n) = \sum_{k=0}^{\infty} h_1(k)\,x(n-k) \tag{3.2}$$

where the * means linear convolution. If all the components in $h_1(n)$ can be represented by some linear combination of orthonormal basis $b_m(n)$, this means that the first-order Volterra kernel $h_1(k)$ in equation 3.2 can be represented by:

$$h_1(k) = \sum_{m=0}^{\infty} a_1(m) b_m(k) \tag{3.3}$$

where $a_l(m)$ are some proper constants. Note that $\{b_m(n), 0 \le m \le \infty\}$ is the set of orthonormal basis, which means:

$$\langle b_l(n), b_m(n) \rangle = \delta(l - m) \tag{3.4}$$

where $\langle \, , \rangle$ denotes the inner product and $\delta(l - m)$ is the Dirac delta function. Substituting equation 3.3 in equation 3.2, we can define the first-order Volterra functional as:

$$Y_1(n) = \sum_{k=0}^{\infty}\sum_{m=0}^{\infty} a_1(m) b_m(k) x(n-k) = \sum_{m=0}^{\infty} a_1(m)[b_m(n) * x(n)] \tag{3.5}$$

Note that equation 3.5 is the first-order homogeneous functional, which means that $Y_1[cx(n)] = cY_1[x(n)]$, where c is a constant. equation 3.5 can be expressed by the block diagram shown in figure 3-2:

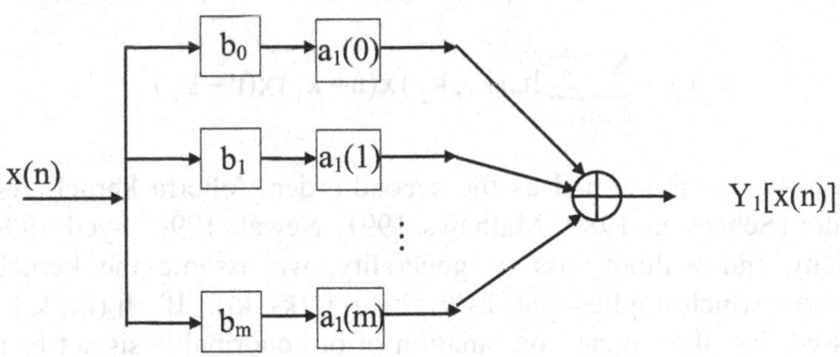

Figure 3-2. First-order Volterra model

For the most general form of first-order Volterra system, we should include the DC term in equation 3.2, which can be expressed in terms of the Volterra functional $Y_0[x(n)]$ and $Y_1[x(n)]$ as:

$$y(n) = h_0 + x(n) * h_1(n) = Y_0[x(n)] + Y_1[x(n)] \tag{3.6}$$

From equation 3.6, we conclude that a general first-order Volterra system with DC term is one for which the response to a linear combination of inputs is the same as the linear combination of the response of each individual input.

3.1.2 Second-Order Volterra Model

The linear combination concept described above can be extended to the second-order case, which is one for which the response to a second-order Volterra system is a linear combination of the individual input signals. Consider the isolated second-order extension version of figure 3-1 shown in figure 3-3:

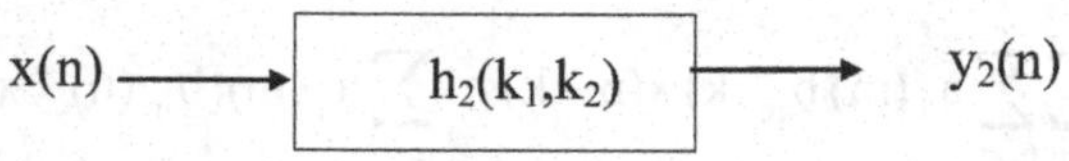

Figure 3-3. Isolated second-order Volterra model block diagram

The response to the input x(n) in figure 3-3 is expressed by $y_2(n)$:

$$y_2(n) = \sum_{k_1=0}^{\infty}\sum_{k_2=0}^{\infty} h_2(k_1,k_2)x(n-k_1)x(n-k_2) \tag{3.7}$$

where $h_2(k_1, k_2)$ is defined as the second-order Volterra kernel. As in the literature (Schetezen 1980, Mathews 1991, Newak 1996, Syed 1994), for simplicity and without loss of generality, we assume the kernel to be symmetric, which implies that $h_2(k_1, k_2) = h_2(k_2, k_1)$. If $h_2(k_1, k_2)$ can be expressed by the linear combination of orthonormal basis set $b_k(n)$, then $h_2(k_1, k_2)$ can be written as

$$h_2(k_1, k_2) = \sum_{m_1=0}^{\infty}\sum_{m_2=0}^{\infty} a_2(m_1,m_2)b_{m_1}(k_1)b_{m_2}(k_2) \tag{3.8}$$

Substituting equation 3.8 in equation 3.7, we can obtain the output as:

$$y_2(n) = \sum_{k_1=0}^{\infty}\sum_{k_2=0}^{\infty} a_2(0,0)\, b_0(k_1)\, b_0(k_2)\, x(n-k_1)\, x(n-k_2)$$

$$+\sum_{k_1=0}^{\infty}\sum_{k_2=0}^{\infty} a_2(1,1)\, b_1(k_1)\, b_1(k_2)\, x(n-k_1)\, x(n-k_2)$$

$$+\ldots+\sum_{k_1=0}^{\infty}\sum_{k_2=0}^{\infty} a_2(1,0)\, b_1(k_1)\, b_0(k_2)\, x(n-k_1)\, x(n-k_2)+\ldots$$

$$= a_2(0,0)[b_0(n) * x(n)]^2 + a_2(1,1)[b_1(n) * x(n)]^2 + \ldots$$

$$+[a_2(0,1)+a_2(1,0)][b_0(n) * x(n)][b_1(n) * x(n)]+\ldots \tag{3.9}$$

equation 3.9, which is defined as the second-order Volterra functional

$$y_2(n) = Y_2[x(n)] \tag{3.10}$$

In equation 3.7, we recognize that all the operations are two-dimensional convolutions, therefore equation 3.10 is a second-order homogeneous functional; i.e., $Y_2[cx(n)] = c^2Y_2[x(n)]$. Consider a special case such that the linear orthonormal set contains two orthonormal bases $\{b_m(n),\ 0 \le m \le 1\}$. From equation 3.9, the second-order Volterra functional $Y_2(n)$ can be expressed as

$$Y_2(n) = a_2(0,0)[b_0(n) * x(n)]^2 + a_2(1,1)[b_1(n) * x(n)]^2$$

$$+[a_2(0,1)+a_2(1,0)][b_0(n) * x(n)][b_1(n) * x(n)] \tag{3.11}$$

The block diagram of equation 3.11 is shown in figure 3-4.

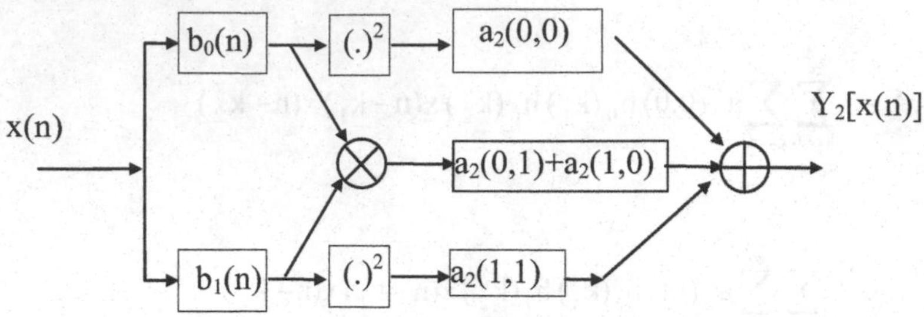

Figure 3-4. Second-order Volterra system with orthonormal basis set $\{b_0, b_1\}$

Based on the description above, the general second-order Volterra system can be represented in terms of $Y_0[x(n)]$, $Y_1[x(n)]$ and $Y_2[x(n)]$ which is

$$y(n) = h_0 + \sum_{k_1=0}^{\infty} h_1(k_1)x(n-k_1) + \sum_{k_1=0}^{\infty}\sum_{k_2=0}^{\infty} h_2(k_1,k_2)x(n-k_1)x(n-k_2)$$

$$= Y_0[x(n)] + Y_1[x(n)] + Y_2[x(n)] \tag{3.12}$$

equation 3.12 can be directly extended to higher-order systems, as shown in the next section.

3.1.3 General Volterra Model

To extend to a higher order, the general Volterra system is considered, where each additive term is the output of a polynomial functional that can be thought of as a multidimensional convolution. The truncated Volterra filter output is the combination of various of these functionals, which can be written as

$$y(n) = Y_0[x(n)] + Y_1[x(n)] + \ldots + Y_j[x(n)] + \ldots = \sum_{i=0}^{\infty} Y_i[x(n)] \tag{3.13}$$

where

$Y_j[x(n)] = \sum_{k_1=0}^{\infty} \ldots \sum_{k_j=0}^{\infty} h_j(k_1,\ldots,k_j)x(n-k_1)\ldots x(n-k_j)$ is the *j*th-order homogeneous functional; i.e., $Y_j[cx(n)] = c\,Y_j[x(n)]$. The *j*th-order Volterra kernel can be expressed in terms of the othonormal basis set as

$$h_j(k_1,\ldots,k_j)$$

$$= \sum_{m_1=0}^{\infty} \ldots \sum_{m_j=0}^{\infty} a_m(m_1,\ldots,m_j)\, b_{m_1}(k_1)\ldots b_{m_j}(k_j) \tag{3.14}$$

where $a_m(m_1,...,m_j)$ are the coefficients. For a Gaussian white input, these functionals in equation 3.13 lack orthogonality in the statistical sense, which means that in general the expected value of any two different Volterra functionals is not equal to zero. This causes two basic difficulties. The first difficulty concerns the measurement of Volterra kernels of a given system, because no exact method of isolating the individual Volterra operator exists. The second problem concerns the large eigenvalue spread issue which implies that slow convergence speed and large misadjustment may be expected, especially for the LMS-type adaptive algorithm.

3.2 Discrete Nonlinear Wiener Representation

Because of these two difficulties in Volterra model representation, with proper rearrangement, a Volterra system can be described in an alternative form which is called the nonlinear Wiener model. The detailed description is presented as follows.

3.2.1 Zeroth- and First-Order Nonlinear Wiener Model

The main problem in the Volterra model is that even when the input signal is Gaussian white noise, the Volterra functionals $Y_j[x(n)]$ lack orthogonality. To overcome this difficulty, instead of using for example Gram-Schmidt orthogonalization procedure, we can rearrange equation 3.6 as:

$$y(n) = g_0[k_0; x(n)] + g_1[k_1, k_{0(1)}; x(n)] \qquad (3.15)$$

where $g_0[k_0; x(n)]$ and $g_1[k_1, k_{0(1)}; x(n)]$ are called zero and first-order non-homogeneous functional respectively. Define $g_0[k_0;x(n)]$ as

$$g_0[k_0; x(n)] = k_0 \qquad (3.16)$$

where k_0 is a constant which is called the zeroth-order kernel. Define $g_1[k_1, k_{0(1)}; x(n)]$ as

$$g_1[k_1, k_{0(1)}; x(n)] = K_1[x(n)] + K_{0(1)}[x(n)] \qquad (3.17)$$

$K_{0(1)}[x(n)]$ and $K_1[x(n)]$ are zeroth- and first-order homogeneous functional with $k_{0(1)}$ and k_1 kernels respectively. In the homogeneous K-functional, the subscript with parenthesis indicates that it is a homogeneous component of the nonhomogeneous g-functional. To develop the orthogonal property, $g_1[k_1, k_{0(1)}; x(n)$ is required to be orthogonal to the homogeneous Volterra functional $Y_0[x(n)]$ (Schetzen 1980). That is:

$$E\{Y_0[x(n)]g_1[k_1, k_{0(1)}; x(n)]\} = 0 \qquad (3.18)$$

To evaluate k_0, we need to substitute equation 3.17 in equation 3.18 which is:

$$E\{Y_0[x(n)]g_1[k_1, k_{0(1)};x(n)]\} = h_0k_{0(1)} + h_0\,E\{\sum_{j=0}^{\infty} k_1(j)\,x(n-j)\} \quad (3.19)$$

Because input x(n) is white noise, $E\{x(n)\} = 0$, thus

$$E\{Y_0[x(n)]g_1[k_1, k_{0(1)};x(n)]\} = h_0k_{0(1)} = 0 \quad (3.20)$$

To satisfy equation 3.20, we obtain $k_{0(1)} = 0$ for any constant h_0.
We define the zero-order and first-order G-functional as

$$G_0[k_0; x(n)] = k_0 \quad (3.21)$$

$$G_1[k_1; x(n)] = g_1[k_1; x(n)] \quad (3.22)$$

From equations 3.21 and 3.22, we can say that *G-functional is the canonical form of g-functional*. Note that, like the g-functional, the G-functional is also a nonhomogeneous functional. For general second-order Volterra systems, therefore, we can have two equivalent Volterra and Wiener model representations which are:

$$y(n) = Y_0[x(n)] + Y_1[x(n)] = G_0[k_0; x(n)] + G_1[k_1; x(n)] \quad (3.23)$$

To explore the orthogonality of G-functionals, we need to evaluate the mean value of any two G-functional mutual products, which for the second-order Volterra system are

$$E\{G_0[k_0; x(n)]G_1[k_1; x(n)]\} = E\{k_0\sum_{j=0}^{\infty} k_{0(1)}(j)\,x(n-j)\} = 0 \quad (3.24a)$$

$$E\{G_0[k_0; x(n)]G_0[k_0; x(n)]\} = k_0^2 \quad (3.24b)$$

$$E\{G_1[k_1;\ x(n)]G_1[k_1;\ x(n)]\} = E\{\sum_{i=0}^{\infty} k_1(i)\,x(n-i)\sum_{j=0}^{\infty} k_1(j)\,x(n-j)\} =$$

$$\sigma_x^2\sum_{i=0}^{\infty} k_1^2(i) \quad (3.24c)$$

Combining equation 3.24a and equation 3.24c, the compact form can be shown as:

$$E\{G_m[k_0; x(n)]G_l[k_1; x(n)]\} = C_m\,\delta(m-l), \quad m, l = 0, 1 \quad (3.25)$$

where $C_0 = k_0^2$ and $C_1 = \sigma_x^2\sum_{i=0}^{\infty} k_1^2(i)$.

3.2.2 Second-Order Nonlinear Wiener Model

To extend the same procedure to second-order systems, we need to rearrange equation 3.12 as:

$$y(n) = g_0[k_0; x(n)] + g_1[k_1, k_{0(1)}; x(n)] + g_2[k_2, k_{1(2)}, k_{0(2)}; x(n)] \quad (3.26)$$

In equation 3.26 $g_2[k_2, k_{1(2)}, k_{0(2)}; x(n)]$ is called second-order non-homogeneous g-functional. Define $g_2[k_2, k_{1(2)}, k_{0(2)}; x(n)]$ as

$$g_2[k_2, k_{1(2)}, k_{0(2)}; x(n)] = K_2[x(n)] + K_{1(2)}[x(n)] + K_{0(2)}[x(n)] \quad (3.27)$$

which is the sum of second-, first-, and zeroth-order homogeneous K-functionals with k_2, $k_{1(2)}$ and $k_{0(2)}$ kernels respectively. To develop the orthogonal property, $g_2[k_2, k_{1(2)}, k_{0(2)}; x(n)]$ is required to be orthogonal with homogeneous Volterra functional $Y_0[x(n)]$ and $Y_1[x(n)]$; that is:

$$E\{Y_0[x(n)]g_2[k_2, k_{1(2)}, k_{0(2)}; x(n)]\} = 0 \quad (3.28a)$$

$$E\{Y_1[x(n)]g_2[k_2, k_{1(2)}, k_{0(2)}; x(n)]\} = 0 \quad (3.28b)$$

To calculate $k_{0(2)}$, we need to substitute equation 3.27 in equation 3.28a, thus:

$$E\{Y_0[x(n)]g_2[k_2, k_{1(2)}, k_{0(2)}; x(n)]\}$$

$$= E\{Y[x(n)]K_2[x(n)]\} + E\{Y[x(n)]K_{1(2)}[x(n)]\}$$

$$+E\{Y[x(n)]K_{0(2)}[x(n)]\}$$

$$= h_0E\{\sum_{j_1=0}^{\infty}\sum_{j_2=0}^{\infty}k_2(j_1, j_2)x(n-j_1)x(n-j_2)\} + h_0E\{\sum_{i=0}^{\infty}k_{2(1)}(i)x(n-i)\}$$

$$+ h_0k_{0(2)} = 0 \quad (3.29)$$

Obviously, the second term of equation 3.28 is equal to zero. Then, $k_{0(2)}$ can be obtained as:

$$k_{0(2)} = -\sigma_x^2\sum_{k_1=0}^{\infty}k_2(k_1, k_1) \quad (3.30)$$

To calculate $k_{1(2)}$, we substitute equation 3.27 in equation 3.28b, therefore:

$$E\{Y_1[x(n)]\, g_2[k_2, k_{1(2)}, k_{0(2)}; x(n)]\}$$

$$= E\{Y_1[x(n)]K_2[x(n)]\} + E\{Y_1[x(n)]K_{1(2)}[x(n)]\} + E\{Y_1[x(n)]K_{0(2)}$$
$$[x(n)]\} = 0 \quad (3.31)$$

The third term in equation 3.31 is equal to zero because:

$$E\{Y_1[x(n)]K_{0(2)}[x(n)]\} = E\{\sum_{k_1=0}^{\infty} h_1(k_1)x(n-k_1)\}k_{0(2)} = 0 \qquad (3.32)$$

The first term in equation 3.31 is equal to:

$$E\{Y_1[x(n)]K_2[x(n)]\} =$$

$$E\{\sum_{k_0=0}^{\infty} h_1(k_0)x(n-k_0)\}\sum_{k_1=0}^{\infty}\sum_{k_2=0}^{\infty} k_2(k_1,k_2)x(n-k_1)x(n-k_2)$$

$$=\sum_{k_0=0}^{\infty}\sum_{k_1=0}^{\infty}\sum_{k_2=0}^{\infty} h_1(k_0)k_2(k_1,k_2)E\{x(n-k_0)x(n-k_1)x(n-k_2)\} = 0$$

$$(3.33)$$

Equation 3.32 is equal to zero because if x(n) is Gaussian white noise, it implies that the mean of the product of an odd number of x's is zero, which means that:

$$E\{x(n-k_0)x(n-k_1)x(n-k_2)\}=0 \qquad (3.34)$$

The second term of equation 3.31 is equal to

$$E\{Y_1[x(n)]K_{1(2)}[x(n)]\} = E\{\sum_{i=0}^{\infty} h_1(i)x(n-i)\}\sum_{j=0}^{\infty} k_{1(2)}(j)x(n-j)\}$$

$$= \sum_{i=0}^{\infty}\sum_{j=0}^{\infty} h_1(i)k_{1(2)}(j)E\{x(n-i)x(n-j)\}$$

$$= \sigma_x^2\sum_{i=0}^{\infty} h_1(i)k_{1(2)}(i) = 0 \qquad (3.35)$$

Because $h_1(i)$ is arbitrary, we can choose $h_1(i) = k_{1(2)}(i)$, then substitute this $h_1(i)$ in equation 3.35 to obtain:

$$\sigma_x^2\sum_{i=0}^{\infty} k_{1(2)}^2(i) = 0 \qquad (3.36)$$

From equation 3.36, we can see that the only choice is $k_{1(2)}(i) = 0$. Now we can define the second-order G_2-functional, which is the canonical form of g_2-functional as:

$$G_2[k_2; x(n)] = g_2[k_2, k_{0(2)}; x(n)]$$

$$= \sum_{j_1=0}^{\infty}\sum_{j_2=0}^{\infty} k_2(j_1,j_2)x(n-j_1)x(n-j_2) - \sigma_x^2\sum_{k_1=0}^{\infty} k_2(k_1,k_1) \qquad (3.37)$$

Therefore, for the general second-order Volterra system, we have both Volterra model and Wiener model representations as

$$y(n) = Y_0[x(n)] + Y_1[x(n)] + Y_2[x(n)]$$

$$= G_0[x(n)] + G_1[k_1; x(n)] + G_2[k_2; x(n)] \tag{3.38}$$

To explore the orthogonality of G-functional, we need to evaluate the mean of any two mutual product values of $G_0[x(n)]$, $G_1[k_1; x(n)]$ and $G_2[k_2; x(n)]$ which for the second-order Volterra system are:

$E\{G_0[k_0; x(n)]\, G_2[k_2; x(n)]\}$

$$= k_0 \sum_{j_1=0}^{\infty}\sum_{j_2=0}^{\infty} k_2(j_1, j_2) E\{x(n-j_1)x(n-j_2)\} - k_0\sigma_x^2 \sum_{k_1=0}^{\infty} k_2(k_1, k_1) = 0 \tag{3.39a}$$

$E\{G_1[k_1; x(n)]\, G_2[k_2; x(n)]\}$

$$= \sum_{i=0}^{\infty}\sum_{j_1=0}^{\infty}\sum_{j_2=0}^{\infty} k_1(i) k_2(j_1, j_2) E\{x(n-i)x(n-j_1)x(n-j_2)\} = 0 \tag{3.39b}$$

$E\{G_2[k_2; x(n)]\, G_2[k_2; x(n)]\}$

$$= E\{ (\sum_{j_1=0}^{\infty}\sum_{j_2=0}^{\infty} k_2(j_1, j_2) x(n-j_1)x(n-j_2) - \sigma_x^2 \sum_{j_3=0}^{\infty} k_2(j_3, j_3))$$

$$(\sum_{i_1=0}^{\infty}\sum_{i_2=0}^{\infty} k_2(i_1, i_2) x(n-i_1)x(n-i_2) \; - \sigma_x^2 \sum_{i_3=0}^{\infty} k_2(i_3, i_3))\}$$

$$= -2\sigma_x^4 \sum_{i_1=0}^{\infty}\sum_{j_3=0}^{\infty} k_2(i_1, i_1) k_2(j_3, j_3) + \sigma_x^4 \sum_{i_3=0}^{\infty}\sum_{j_3=0}^{\infty} k_2(i_3, i_3) k_2(j_3, j_3)$$

$$= \sum_{i_1=0}^{\infty}\sum_{i_2=0}^{\infty}\sum_{j_1=0}^{\infty}\sum_{j_2=0}^{\infty} k_2(i_1, i_2) k_2(j_1, j_2) E\{x(n-i_1)x(n-i_2)\} E\{x(n-j_1)x(n-j_2)\}$$

$+$

$$\sum_{i_1=0}^{\infty}\sum_{i_2=0}^{\infty}\sum_{j_1=0}^{\infty}\sum_{j_2=0}^{\infty} k_2(i_1, i_2) k_2(j_1, j_2) E\{x(n-i_1)x(n-j_1)\} E\{x(n-i_2)x(n-j_2)\}$$

$+$

$$\sum_{i_1=0}^{\infty}\sum_{i_2=0}^{\infty}\sum_{j_1=0}^{\infty}\sum_{j_2=0}^{\infty} k_2(i_1,i_2)k_2(j_1,j_2)E\{x(n-i_1)x(n-j_2)\}E\{x(n-j_1)x(n-i_2)\}$$

$$-2\sigma_x^4\sum_{i_1=0}^{\infty}\sum_{j_3=0}^{\infty} k_2(i_1,i_1)k_2(j_3,j_3)+\sigma_x^4\sum_{i_3=0}^{\infty}\sum_{j_3=0}^{\infty} k_2(i_3,i_3)k_2(j_3,j_3)$$

$$=\sigma_x^4\sum_{i_1=0}^{\infty}\sum_{j_2=0}^{\infty} k_2(i_1,i_1)k_2(j_2,j_2)$$

$$+2\sigma_x^4\sum_{i_1=0}^{\infty}\sum_{i_2=0}^{\infty} k_2(i_1,i_2)k_2(i_1,i_2)-2\sigma_x^4\sum_{i_1=0}^{\infty}\sum_{j_3=0}^{\infty} k_2(i_1,i_1)k_2(j_3,j_3)$$

$$+\sigma_x^4\sum_{i_3=0}^{\infty}\sum_{j_3=0}^{\infty} k_2(i_3,i_3)k_2(j_3,j_3)$$

$$=2\sigma_x^4\sum_{i_1=0}^{\infty}\sum_{i_2=0}^{\infty} k_2(i_1,i_2)k_2(i_1,i_2)=2!(\sigma_x^2)^2\sum_{i_1=0}^{\infty}\sum_{i_2=0}^{\infty} k_2(i_1,i_2)k_2(i_1,i_2) \tag{3.39c}$$

Combining equations 3.39a and 3.39c, we have

$$E\{G_m[k_0;\, x(n)]G_l[k_1;\, x(n)]\} = C_m\delta(m-l),\quad m,l = 0,1,2 \tag{3.40}$$

where $C_2 = 2!(\sigma_x^2)^2\sum_{i_1=0}^{\infty}\sum_{i_2=0}^{\infty} k_2(i_1,i_2)k_2(i_1,i_2)$.

This confirms the orthogonality of the second-order G-functional.

3.2.3 Third-Order Nonlinear Wiener Model

To have a clearer insight into the Wiener model before we go to the general higher-order expressions, let us consider a third-order model in this section. Assume:

$$\begin{aligned} y(n) = {} & g_0[k_0;\, x(n)] + g_1[k_1, k_{0(1)};\, x(n)] + g_2[k_2, k_{1(2)}, k_{0(2)};\, x(n)] \\ & + g_3[k_3, k_{2(3)}, k_{1(3)}, k_{0(3)};\, x(n)] \end{aligned} \tag{3.41}$$

where $g_3[k_3, k_{2(3)}, k_{1(3)}, k_{0(3)};\, x(n)]$ is called third-degree non-homogeneous functional. Define $g_3[k_3, k_{2(3)}, k_{1(3)}, k_{0(3)};\, x(n)]$ as:

$$g_3[k_3,\ k_{2(3)},\ k_{1(3)},\ k_{0(3)};\ x(n)] = K_3\ [x(n)] + K_{2(3)}\ [x(n)] + K_{1(3)}[x(n)] + K_{0(3)}[x(n)] \quad (3.42)$$

Equation 3.42 is the sum of the third-, second-, first-, and zeroth-order homogeneous functional with k_3, $k_{2(3)}$, $k_{1(3)}$ and $k_{0(3)}$ kernel respectively. Similarly, to develop the orthogonal property, $g_3[k_3, k_{2(3)}, k_{1(3)}, k_{0(3)}; x(n)]$ is required to be an orthogonal homogeneous Volterra functional $Y_0[x(n)]$, $Y_1[x(n)]$ and $Y_2[x(n)]$; that is:

$$E\{Y_0[x(n)]g_3[k_3, k_{2(3)}, k_{1(3)}, k_{0(3)}; x(n)]\} = 0 \quad (3.43a)$$

$$E\{Y_1[x(n)]g_3[k_3, k_{2(3)}, k_{1(3)}, k_{0(3)}; x(n)]\} = 0 \quad (3.43b)$$

$$E\{Y_2[x(n)]g_3[k_3, k_{2(3)}, k_{1(3)}, k_{0(3)}; x(n)]\} = 0 \quad (3.43c)$$

To calculate $k_{0(3)}$, we need to substitute equation 3.42 in equation 3.43a, thus we have:

$$E\{Y_0[x(n)]g_3[k_3, k_{2(3)}, k_{1(3)}, k_{0(3)}; x(n)]\}$$

$$= E\{Y_0[x(n)]K_3[x(n)]\} + E\{Y_0[x(n)]K_{2(3)}[x(n)]\} + E\{Y_0[x(n)]K_{1(3)}[x(n)]\}$$

$$+ E\{Y_0[x(n)]K_{0(3)}[x(n)]\}$$

$$= h_0 E\{\sum_{k_1=0}^{\infty}\sum_{k_2=0}^{\infty}\sum_{k_3=0}^{\infty} k_3(k_1,k_2,k_3)x(n-k_1)x(n-k_2)x(n-k_3)\}$$

$$+ h_0 E\{\sum_{k_1=0}^{\infty}\sum_{k_2=0}^{\infty} k_{2(3)}(k_1,k_2)x(n-k_1)x(n-k_2)\}$$

$$+ h_0 E\{\sum_{m=0}^{\infty} k_{1(3)}(k)x(n-k)\} + h_0 k_{3(0)}$$

$$= h_0 E\{\sum_{k_1=0}^{\infty}\sum_{k_2=0}^{\infty} k_{2(3)}(k_1,k_2)x(n-k_1)x(n-k_2)\} + h_0 k_{3(0)} = 0 \quad (3.44)$$

To satisfy equation 3.44, we can find that $k_{3(0)}$ is equal to:

$$k_{0(3)} = -\sigma_x^2 \sum_{k_1=0}^{\infty} k_{2(3)}(k_1,k_1) \quad (3.45)$$

To calculate $k_{1(3)}$, we need to substitute equation 3.42 in equation 3.43b, and we have:

$$E\{Y_1[x(n)]g_3[k_3, k_{2(3)}, k_{1(3)}, k_{0(3)}; x(n)]\} =$$

$$E\{Y_1[x(n)]K_3[x(n)]\}+E\{Y_1[x(n)]K_{2(3)}[x(n)]\}+E\{Y_1[x(n)]K_{1(3)}[x(n)]\} +E\{Y_1[x(n)]K_{0(3)}[x(n)]\} \quad (3.46)$$

The second term and fourth term in equation 3.46 are equal to zero because these two terms involve the odd-numbered of the terms whose mean value is zero if x(n) is white Gaussian noise . Thus:

$$E\{Y_1[x(n)]g_3[k_3, k_{2(3)}, k_{1(3)}, k_{0(3)}; x(n)]\} = E\{Y_1[x(n)]K_{1(3)}[x(n)]\} +$$

$$E\{Y_1[x(n)]K_{(3)}[x(n)]\} =$$

$$E\{\sum_{k_0=0}^{\infty} h_1(k_0)x(n-k_0)\sum_{k_1=0}^{\infty} k_{1(3)}(k_1)x(n-k_1)\} +$$

$$E\{\sum_{k_0=0}^{\infty} h_1(k_0)x(n-k_0)$$

$$\sum_{k_1=0}^{\infty}\sum_{k_2=0}^{\infty}\sum_{k_3=0}^{\infty} k_3(k_1,k_2,k_3)x(n-k_1)x(n-k_2)x(n-k_3)\} =$$

$$\sigma_x^2\sum_{k_0=0}^{\infty} h_1(k_0)\,k_{1(3)}(k_0) + 3\sigma_x^4\sum_{k_0=0}^{\infty}\sum_{k_1=0}^{\infty} h_1(k_0)\,k_3(k_0,k_1,k_1) =$$

$$\sum_{k_0=0}^{\infty} h_1(k_0)\left(\sigma_x^2\, k_{1(3)}(k_0) + 3\sigma_x^4\sum_{k_1=0}^{\infty} k_3(k_0,k_1,k_1)\right) = 0 \quad (3.47)$$

For any $h_1(k_0)$ kernel, the term in parenthesis should be equal to zero, which implies we can express $k_{1(3)}$ as:

$$k_{1(3)}(k_0) = -3\sigma_x^2\sum_{k_1=0}^{\infty} k_3(k_0,k_1,k_1) \quad (3.48)$$

To calculate $k_{2(3)}$, we need to substitute equation 3.42 in equation 3.43c, and we have:

$$E\{Y_2[x(n)]g_3[k_3, k_{2(3)}, k_{1(3)}, k_{0(3)}; x(n)]\}=$$

$$E\{Y_2[x(n)]K_{0(3)}[x(n)]\}+E\{Y_2[x(n)]K_{1(3)}[x(n)]\}+E\{Y_2[x(n)]K_{2(3)}[x(n)]\}+E\{Y_2[x(n)]K_{(3)}[x(n)]\} \quad (3.49)$$

The second term and fourth term in equation 3.43 are equal to zero because each involves the odd number of the terms production whose mean values is equal to zero. Therefore:

$$E\{Y_2[x(n)]g_3[k_3, k_{2(3)}, k_{1(3)}, k_{0(3)}; x(n)]\}$$

$$= E\{Y_2[x(n)]K_{0(3)}[x(n)]\} + E\{Y_2[x(n)]K_{2(3)}[x(n)]\} \quad (3.50)$$

The first term of equation 3.50 is:

$$E\{Y_2[x(n)]K_{0(3)}[x(n)]\}$$

$$= E\{k_{0(3)}\sum_{k_0=0}^{\infty}\sum_{k_1=0}^{\infty}h_2(k_0,k_1)x(n-k_0)x(n-k_1)\}$$

$$= \sigma_x^2\, k_{0(3)}\sum_{k_0=0}^{\infty}h_2(k_0,k_0) \quad (3.51)$$

The second term of equation 3.50 is:

$$E\{Y_2[x(n)]K_{2(3)}[x(n)]\}$$

$$= E\{\sum_{k_0=0}^{\infty}\sum_{k_1=0}^{\infty}h_2(k_0,k_1)x(n-k_0)x(n-k_1)\sum_{j_0=0}^{\infty}\sum_{j_1=0}^{\infty}k_{2(3)}(j_0,j_1)x(n-j_0)x(n-j_1)\}$$

$$= \sum_{k_0=0}^{\infty}\sum_{k_1=0}^{\infty}\sum_{j_0=0}^{\infty}\sum_{j_1=0}^{\infty}h_2(k_0,k_1)k_{2(3)}(j_0,j_1)E\{x(n-k_0)x(n-k_1)x(n-j_0)x(n-j_1)\}$$

$$= \sum_{k_0=0}^{\infty}\sum_{k_1=0}^{\infty}\sum_{j_0=0}^{\infty}\sum_{j_1=0}^{\infty}h_2(k_0,k_1)k_{2(3)}(j_0,j_1)E\{x(n-k_0)x(n-k_1)\}E\{x(n-j_0)x(n-j_1)\}$$

$$+ \sum_{k_0=0}^{\infty}\sum_{k_1=0}^{\infty}\sum_{j_0=0}^{\infty}\sum_{j_1=0}^{\infty}h_2(k_0,k_1)k_{2(3)}(j_0,j_1)E\{x(n-k_0)x(n-j_0)\}E\{x(n-k_1)x(n-j_1)\}$$

$$+\sum_{k_0=0}^{\infty}\sum_{k_1=0}^{\infty}\sum_{j_0=0}^{\infty}\sum_{j_1=0}^{\infty}h_2(k_0,k_1)k_{2(3)}(j_0,j_1)E\{x(n-k_0)x(n-j_1)\}E\{x(n-j_0)x(n-k_1)\}$$

$$=\sigma_x^4\sum_{k_0=0}^{\infty}\sum_{j_0=0}^{\infty}h_2(k_0,k_0)k_{2(3)}(j_0,j_0)+\sigma_x^4\sum_{j_0=0}^{\infty}\sum_{j_1=0}^{\infty}h_2(j_0,j_1)k_{2(3)}(j_0,j_1)$$

$$+\sigma_x^4\sum_{j_0=0}^{\infty}\sum_{j_1=0}^{\infty}h_2(j_1,j_0)k_{2(3)}(j_0,j_1)$$

$$=\sigma_x^4\sum_{k_0=0}^{\infty}\sum_{j_0=0}^{\infty}h_2(k_0,k_0)k_{2(3)}(j_0,j_0)+2\sigma_x^4\sum_{j_0=0}^{\infty}\sum_{j_1=0}^{\infty}h_2(j_0,j_1)k_{2(3)}(j_0,j_1) \quad (3.52)$$

Substituting equation 3.51 and equation 3.52 in equation 3.50, we have:

$$E\{Y_2[x(n)]g_3[k_3, k_{2(3)}, k_{1(3)}, k_{0(3)}; x(n)]\} \quad (3.53)$$

$$=\sigma_x^2 k_{0(3)}\sum_{k_0=0}^{\infty}h_2(k_0,k_0)+\sigma_x^4\sum_{k_0=0}^{\infty}\sum_{j_0=0}^{\infty}h_2(k_0,k_0)k_{2(3)}(j_0,j_0)$$
$$+2\sigma_x^4\sum_{j_0=0}^{\infty}\sum_{j_1=0}^{\infty}h_2(j_0,j_1)k_{2(3)}(j_0,j_1)$$

Substitute equation 4.45 in equation 4.53; the first term and second term cancel each other, and only the third term is left. Thus:

$$E\{Y_2[x(n)]g_3[k_3, k_{2(3)}, k_{1(3)}, k_{0(3)}; x(n)]\}$$

$$=2\sigma_x^4\sum_{j_0=0}^{\infty}\sum_{j_1=0}^{\infty}h_2(j_0,j_1)k_{2(3)}(j_0,j_1)=0 \quad (3.54)$$

For arbitrary $h_2(k_0, k_1)$,

$$k_{2(3)}(j_0, j_1) = 0 \quad (3.55)$$

is the only choice to satisfy equation 3.54. Then, from equation 3.45, we know immediately that

$$k_{0(3)} = 0. \quad (3.56)$$

Therefore, the canonical form of equation 3.42 is expressed by a third-order G-functional as:

$$G_3[k_3; x(n)] = g_3[k_3, k_{1(3)}; x(n)]$$

$$= \sum_{k_1=0}^{\infty}\sum_{k_2=0}^{\infty}\sum_{k_3=0}^{\infty} k_3(k_1, k_2, k_3)\, x(n-k_1)\, x(n-k_2)\, x(n-k_3)$$

$$+ \sum_{k_1=0}^{\infty} k_{1(3)}(k_1)\, x(n-k_1) \tag{3.57}$$

For a general third-order Volterra system, we can have both Volterra mode representation and equivalent Wiener model representation respectively as:

$$y(n) = Y_0[x(n)] + Y_1[x(n)] + Y_2[x(n)] + Y_3[x(n)]$$

$$= G_0[x(n)] + G_1[k_1; x(n)] + G_2[k_2; x(n)] + G_3[k_3; x(n)] \tag{3.58}$$

To explore the orthogonality of the G-functional, we need to evaluate the mutual product mean value of $G_0[x(n)]$, $G_1[k_1; x(n)]$, $G_2[k_2; x(n)]$, and $G_3[k_3; x(n)]$. As before, we obtain:

$$E\{G_0[k_0; x(n)]\, G_3[k_3; x(n)]\} = 0 \tag{3.59a}$$

$$E\{G_1[k_0; x(n)]\, G_3[k_3; x(n)]\} = 0 \tag{3.59b}$$

$$E\{G_2[k_0; x(n)]\, G_3[k_3; x(n)]\} = 0 \tag{3.59c}$$

It is interesting to see the results of $E\{G_3[k_3; x(n)]\, G_3[k_3; x(n)]\}$:

$$E\{G_3[k_3; x(n)]\, G_3[k_3; x(n)]\}$$

$$= E\{G_3[k_3; x(n)]\, (K_3[x(n)] + K_{1(3)}[x(n)])\}$$

$$= E\{G_3[k_3; x(n)] K_3[x(n)]\} + E\{G_3[k_3; x(n)] K_{1(3)}[x(n)]\} \tag{3.60}$$

The second term of equation 3.60 is equal to zero because the third-order G-functional is orthogonal to the homogeneous functional whose order is less than three. The first term of equation 3.60 can be expanded as:

$$E\{G_3[k_3; x(n)] K_3[x(n)]\}$$

$$= \sum_{i_1=0}^{\infty}\sum_{i_2=0}^{\infty}\sum_{i_3=0}^{\infty}\sum_{j_1=0}^{\infty}\sum_{j_2=0}^{\infty}\sum_{j_3=0}^{\infty} k_3(i_1,i_2,i_3) k_3(j_1,j_2,j_3) E\{x(n-i_1)x(n-i_2)x(n-i_3)x(n-j_1)x(n-j_2)x(n-j_3)\}$$

$$+\sum_{j_1=0}^{\infty}\sum_{j_2=0}^{\infty}\sum_{j_3=0}^{\infty}\sum_{i_4=0}^{\infty} k_3(j_1,j_2,j_3)k_{1(3)}(i_4)E\{x(n-j_1)x(n-j_2)x(n-j_3)x(n-i_4)\} \quad (3.61)$$

By expanding the expectation value of six Gaussian random variables, the first term of equation 3.61 can be expressed as:

$$3!(\sigma_x^2)^3\sum_{i_1=0}^{\infty}\sum_{i_2=0}^{\infty}\sum_{i_3=0}^{\infty} k_3(i_1,i_2,i_3)k_3(i_1,i_2,i_3) +$$

$$3(\sigma_x^2)^3\sum_{i_1=0}^{\infty}\sum_{i_3=0}^{\infty}\sum_{j_2=0}^{\infty} k_3(i_1,i_1,i_3)k_3(i_3,j_2,j_2) +$$

$$3(\sigma_x^2)^3\sum_{i_1=0}^{\infty}\sum_{i_2=0}^{\infty}\sum_{j2=0}^{\infty} k_3(i_1,i_2,i_1)k_3(i_2,j_2,j_2) +$$

$$3(\sigma_x^2)^3\sum_{i_1=0}^{\infty}\sum_{i_2=0}^{\infty}\sum_{j_2=0}^{\infty} k_3(i_1,i_2,i_2)k_3(i_1,j_2,j_2) \quad (3.62a)$$

Expanding the expectation value of four Gaussian random variables, the second term of equation 3.61 can be expressed as:

$$-3(\sigma_x^2)^3\sum_{j_1=0}^{\infty}\sum_{j_2=0}^{\infty}\sum_{j_2=0}^{\infty} k_3(j_1,j_1,j_3)k_3(j_3,i_2,i_2)$$

$$-3(\sigma_x^2)^3\sum_{j_1=0}^{\infty}\sum_{j_2=0}^{\infty}\sum_{i_3=0}^{\infty} k_3(j_1,j_2,j_1)k_3(j_2,i_3,i_3)$$

$$-3(\sigma_x^2)^3\sum_{j_1=0}^{\infty}\sum_{j_3=0}^{\infty}\sum_{i_3=0}^{\infty} k_3(j_1,j_2,j_2)k_3(j_1,i_3,i_3) \quad (3.62b)$$

Substituting equation 3.62a and equation 3.62b in equation 3.61 yields:

$E\{G_3[k_3; x(n)]K_3[x(n)]\} =$

$$3!(\sigma_x^2)^3\sum_{i_1=0}^{\infty}\sum_{i_2=0}^{\infty}\sum_{i_3=0}^{\infty} k_3(i_1,i_2,i_3)k_3(i_1,i_2,i_3) \quad (3.63)$$

This implies that the autocorrelation of third-order G-functional is:

$$E\{G_m[k_m; x(n)]G_l[k_l; x(n)]\} = C_3\delta(m-l), \quad m,l = 0,1,2,3 \quad (3.64)$$

where $C_3 = 3!(\sigma_x^2)^3 \sum_{i_1=0}^{\infty}\sum_{i_2=0}^{\infty}\sum_{i_3=0}^{\infty} k_3(i_1,i_2,i_3)k_3(i_1,i_2,i_3)$.

Therefore, the orthogonality is confirmed.

Before leaving this section, a summary of the Wiener kernels and autocorrelation of G-functionals for zeroth- to third-order are shown in table 3-1. From table 3-1, we note that there exist some relationships between G-functional and g-functional. For instance:

$$G_3[k_3; x(n)] = g_3[k_3, k_{1(3)}; x(n)] \tag{3.65a}$$

$$G_2[k_2; x(n)] = g_2[k_2, k_{0(2)}; x(n)] \tag{3.65b}$$

And we note that, for each order, the Wiener kernel is related to the leading kernel, which means that all the Wiener kernels can be expressed in terms of the leading kernel. This is also true for the autocorrelation of G-functional. The expression in table 3-1 can be extended to the general higher order conditions which will be shown in the next section. In appendix 3A, there is one numerical example which can helps to better explain the relationship of all the kernels.

Table 3-1. Kernel of Wiener model and autocorrelation of G-functional for 0th- to 3rd-order

	Kernel of Wiener model	Autocorrelation of G-functional
0th-order	k_0 = Leading kernel of 0th-order	$C_0 = k_0^2$
1st-order	k_1 = Leading kernel of 1st-order $k_{(0)1} = 0$	$C_1 = \sigma_x^2 \sum_{i_1=0}^{\infty} k_1^2(i_1)$
2nd-order	k_2 = Leading kernel of 2nd-order $k_{1(2)} = 0$ $k_{0(2)} = -\sigma_x^2 \sum_{k_1=0}^{\infty} k_2(k_1,k_1)$	$C_2 =$ $2!(\sigma_x^2)^2$ $\sum_{i_1=0}^{\infty}\sum_{i_2=0}^{\infty} k_2(i_1,i_2)k_2(i_1,i_2)$
3rd-order	k_3 = Leading kernel of 3rd-order $k_{2(3)} = 0$ $k_{1(3)} = -3\sigma_x^2 \sum_{k_1=0}^{\infty} k_3(k_0,k_1,k_1)$ $k_{0(3)} = 0$	$C_3 =$ $3!(\sigma_x^2)^3$ $\sum_{i_1=0}^{\infty}\sum_{i_2=0}^{\infty}\sum_{i_3=0}^{\infty} k_3(i_1,i_2,i_3)k_3(i_1,i_2,i_3$

3.2.4 General Nonlinear Wiener Model

To extend this result to the *j*th-order, the general Volterra system is considered. The general Volterra system may have equivalent non-homogeneous G-functional representation as:

$$y(n) = G_0[k_0; x(n)] + G_1[k_1; x(n)] + \ldots + G_j[k_j; x(n)] + \ldots$$

$$= \sum_{i=0}^{\infty} G_i[k_i; x(n)] \tag{3.66}$$

From the previous section, the general relation between the non-homogeneous G-functional and the homogeneous Wiener kernel can be deduced as:

For m = even,

$$G_m[k_m; x(n)] = g_m[k_m, k_{m-2(m)}, k_{m-4(m)}, \ldots, k_{0(m)}; x(n)] \tag{3.67a}$$

For m = odd,

$$G_m[k_m; x(n)] = g_m[k_m, k_{m-2(m)}, k_{m-4(m)}, \ldots, k_{1(m)}; x(n)] \tag{3.67b}$$

Note that, for m = even, all the homogeneous functional kernels with odd index numbers are equal to zero. For m = odd, all the homogeneous functional kernels with even index numbers are equal to zero. The more general expression of equation 3.67a and equation 3.67b can be expressed as a function of the Wiener kernel as:

$G_m[k_m; x(n)] =$

$$\sum_{r=0}^{[m/2]} \sum_{i_1=0}^{\infty} \ldots \sum_{i_{m-2r}=0}^{\infty} k_{m-2r(m)}(i_1, i_2, \ldots, i_{m-2r}) x(n-i_1) \ldots x(n-i_{m-2r}) \tag{3.68a}$$

where [m/2] means the largest integer less than or equal to m/2 and

$$k_{m-2r(m)}(i_1, i_2, \ldots, i_{m-2r}) = \frac{(-1)^r m! (\sigma_x^2)^r}{(m-2r)! r! 2^r} \sum_{j_1=0}^{\infty} \ldots \sum_{j_r=0}^{\infty} k_m(\underbrace{j_1, j_1, \ldots, j_r, j_r}_{2r\text{'s}}, \underbrace{i_1, i_2, \ldots, i_{m-2r}}_{(m-2r)\text{'s}}) \tag{3.68b}$$

The detailed derivation of equation 3.68b is shown in appendix 3B. From equation 3.62a, we note that all the homogeneous functional kernels $k_{m-2r(m)}$ can be expressed in terms of the leading kernel k_m. Furthermore, the Wiener G-functional series is an orthogonal series for white Gaussian inputs, that is:

$$E\{G_l[k_l\,;x(n)]G_m[k_m\,;x(n)]\} = C_m\delta(l-m) \quad l, m = 0, 1, 2, ..., p \qquad (3.69)$$

where $C_m = m!(\sigma_x^2)^m \sum_{i_1=0}^{\infty}\sum_{i_2=0}^{\infty}...\sum_{i_m=0}^{\infty} k_m(i_1,i_2,...,i_m)k_m(i_1,i_2,...,i_m)$.

Equation 3.69 can be obtained by direct inspection from table 3-1. The derivation procedure is exactly the same as in previous sections.

For the general Volterra system, which is described by the Volterra series, we now have two equivalent representations: the Volterra model and the Wiener model. This means that:

$$y(n) = \sum_{i=0}^{\infty} Y_i[x(n)] = \sum_{i=0}^{\infty} G_i[k_i; x(n)] \qquad (3.70)$$

To explore the relationship of Volterra kernels and Wiener kernels, we need to expand equation 3.70 and compare the kernels for both the Volterra model and the Wiener model. This relationship is summarized in table 3-2 (Schetzen 1980, Hashad 1994).

Table 3-2. The kernel relation between Volterra and Wiener model of mth-order for m = even number

	h_0	h_1	h_2	h_3	...	h_{m-2}	h_{m-1}	h_m
G_m	$k_{0(m)}$		$k_{2(m)}$			$k_{m-2(m)}$		k_m
G_{m-1}		$k_{1(m-1)}$		$k_{3(m-1)}$			k_{m-1}	
G_{m-2}	$k_{0(m-2)}$		$k_{2(m-2)}$			k_{m-2}		
...				...				
G_2	$k_{0(2)}$		k_2					
G_1		k_1						
G_0	k_0							

From table 3-2, we can show that the relationship between Volterra kernels and Wiener kernels is:

$$
\begin{aligned}
h_m &= k_m \\
h_{m-1} &= k_{m-1} \\
h_{m-2} &= k_{m-2} + k_{m-2(m)} \\
&\dots \\
h_1 &= k_1 + k_{1(3)} + \cdots + k_{1(m-3)} + k_{1(m-1)} \\
h_0 &= k_0 + k_{0(2)} + \cdots + k_{0(m-2)} + k_{0(m)}
\end{aligned} \tag{3.71}
$$

In equation 3.71, we obtain one unique Volterra kernel from Wiener kernels. In other words, the Volterra series is a subset of the Wiener G-functional representation. This means that any system that can be described by a Volterra series also can have the Wiener G-functional representation. However, from the adaptive filtering point of view, a Volterra system may be identified more efficiently by a Wiener representation than by a Volterra series because of convergence problems which will be discussed in the next chapter.

3.3 Detailed Nonlinear Wiener Model Representation

As in equation 3.14, the homogeneous kernel k_m can be approximated by the same set of orthonormal bases as:

$$k_m(i_1, i_2, \dots, i_m) = \sum_{n_1=0}^{\infty} \dots \sum_{n_m=0}^{\infty} c_m(n_1, \dots, n_m)\, b_{n_1}(i_1) \dots b_{n_m}(i_m) \tag{3.72}$$

For simplicity, we drop the parenthesis indices in k_m and b_{m_i} from now on. Substituting equation 3.72 in equation 3.68, $G_m[k_m; x(n)]$ can be shown as:

$$G_m[k_m; x(n)] = G_m[\sum_{n_1=0}^{\infty} \dots \sum_{n_m=0}^{\infty} c_m(n_1, \dots, n_m)\, b_{n_1} \dots b_{n_m}; x(n)]$$

$$= \sum_{n_1=0}^{\infty} \dots \sum_{n_m=0}^{\infty} c_m(n_1, \dots, n_m)\, G_m[b_{n_1} \dots b_{n_m}; x(n)] \tag{3.73}$$

To see more detailed properties of $G_m[k_m; x(n)]$, the equivalent series of equation 3.73 can be expressed as:

$$G_m[k_m; x(n)] = c_m(0,\ldots,0)\, G_m[b_0\ldots b_0; x(n)] +$$

$$c_m(0,\ldots,0,1) G_m[b_0\ldots b_0 b_1; x(n)] + \ldots +$$

$$c_m(\underbrace{m_1,\ldots,m_1}_{k_1\text{'s}},\underbrace{m_2,\ldots,m_2}_{k_2\text{'s}},\ldots,\underbrace{n_M,\ldots,n_M}_{k_M\text{'s}})\, G_m[$$

$$\underbrace{b_{m_1}\ldots b_{m_1}}_{k_1\text{'s}}\underbrace{b_{m_2}\ldots b_{m_2}}_{k_2\text{'s}}\ldots\underbrace{b_{m_M}\ldots b_{m_M}}_{k_M\text{'s}}; x(n)] + \ldots$$

$$= c_m(0,\ldots,0)\, G_m[b_0^{(m)}; x(n)] + c_m(0,\ldots,0,1) G_m[b_0^{(m-1)} b_1; x(n)] + \ldots$$

$$+\, c_m(\underbrace{m_1,\ldots,m_1}_{k_1\text{'s}},\underbrace{m_2,\ldots,m_2}_{k_2\text{'s}},\ldots,\underbrace{n_M,\ldots,n_M}_{k_M\text{'s}})\, G_m[b_{m_1}^{(k_1)} b_{m_2}^{(k_2)}\ldots b_{m_M}^{(k_M)};$$

$$x(n)] + \ldots$$

$$= \sum_{m_1=0}^{\infty}\ldots\sum_{m_m=0}^{\infty} c_m(n_1,\ldots,n_m)\, G_m[b_{m_1}^{(k_1)}\ldots b_{m_M}^{(k_M)}; x(n)] \tag{3.74}$$

where $\sum_{i=1}^{M} k_i = m$, M is the number of different bases.

The $b_{m_i}^{(k_i)}$ denotes that the orthonormal bases b_{m_i} appears k_i times. From appendix 3B, we note that the G-functional is separable, therefore:

$$G_m[b_{m_1}^{(k_1)} b_{m_2}^{(k_2)}\ldots b_{m_M}^{(k_M)}; x(n)]$$

$$= G_{k_1}[b_{m_1}^{(k_1)}; x(n)]\, G_{k_2}[b_{m_2}^{(k_2)}; x(n)] \ldots G_{k_M}[b_{m_M}^{(k_M)}; x(n)] \tag{3.75}$$

Then, equation 3.74 can be rewritten as:

$$G_m[k_m; x(n)] = \sum_{n_1=0}^{\infty}\ldots\sum_{n_m=0}^{\infty} c_m(n_1,\ldots,n_m) \prod_{i=1}^{M} G_{k_i}[b_{m_i}^{(k_i)}; x(n)] \tag{3.76}$$

For a white Gaussian input of σ_x^2 variance, $G_{k_i}[b_{m_i}^{(k_i)}; x(n)]$ can be found

by substituting equation 3.76 in equation 3.68, thus:

$$G_{k_i}[b_{m_i}^{(k_i)}; x(n)] = \sum_{m=0}^{[k_i/2]} \frac{(-1)^m k_i! (\sigma_x^2)^m}{(k_i - 2m)! k_i! 2^m} \sum_{i_1=0}^{\infty}\ldots\sum_{i_{m-2\tau}=0}^{\infty}$$

$$\sum_{j_1=0}^{\infty}\cdots\sum_{j_r=0}^{\infty} b_1(j_1)b_1(j_1),\ldots,b_r(j_r),b_r(j_r),b_{m_i}(i_1),\ldots,b_{m_i}(i_{k_i-2r})$$

$$x(n-i_1)\ldots x(n-i_{k_i-2r}) = \sum_{m=0}^{[k_i/2]}\frac{(-1)^m k_i!}{m!(k_i-2m)!}\left(\frac{\sigma_x^2}{2}\right)^m \left(b_{m_i}(n) * x(n)\right)^{k_i-2m}$$

(3.77)

We can get equation 3.77 by using the fact that

$$\sum_{j_i=0}^{\infty} b_i^2(j_i) = 1.$$

Comparing with the *P*th-order Hermite polynomial (Schetzen 1980):

$$H_p[z_i(n)] = \sum_{m=0}^{[P/2]}\frac{(-1)^m p!}{m!(p-2m)!}\left(\frac{A}{2}\right)^m z_i^{P-2m}(n) \tag{3.78}$$

where A is a constant. We note that the G-functional and the Hermite polynomial have the same form, which means that the G-functional can have the same properties as the Hermite polynomial. We also note that if $z_i(n)$ is a white Gaussian noise, the Hermite polynomial has zero mean and satisfies the orthogonal properties which are:

$$E\{H_{p,p\neq 0}(z)\} = 0 \tag{3.79a}$$

$$E\{H_p(z_i)H_q(z_j)\} = p!A^p\delta(i-j)\delta(p-q) \tag{3.79b}$$

Note that $H_0(z) = 1$. Comparing equation 3.71 and equation 3.72, we know immediately that:

$$G_{k_i}[b_{m_i}^{(k_i)}; x(n)] = H_{k_i}[z_{m_i}(n)] \tag{3.80}$$

where $z_{m_i}(n) = b_{m_i}(n) * x(n)$. Substitute equation 3.80 into equation 3.76. Then $G_m[k_m; x(n)]$ can be rewritten as

$$G_m[k_m; x(n)] = \sum_{n_1=0}^{\infty}\cdots\sum_{n_m=0}^{\infty} c_m(n_1,\ldots,n_m)\prod_{i=1}^{M} H_{k_i}[z_{m_i}(n)] \tag{3.81}$$

Define the product term of equation 3.81 as $\widetilde{Q}$-polynomial

$$\tilde{Q}^{(m)}_{\alpha_m}(n) = \prod_{i=1}^{M} H_{k_i}[z_{m_i}(n)] \tag{3.82}$$

The superscript m indicates the m-th degree $\tilde{Q}$-polynomial and $\alpha_m = \underbrace{n_1, n_2, \ldots, n_m}_{m's}\big|_{m=1,\ldots,M}$, which can be obtained by the permutation of the elements of n_1, n_2, ..., n_m. Recall that M is a constant which indicates the number of different H_{k_j} used in equation 3.82, and $\sum_{i=1}^{M} k_i = m$. A simple example may be helpful to get a clearer insight into equation 3.76. If m = 3 (3rd-order case), there are three indices n_1, n_2, and n_3. The three different cases, equation 3.82, are:

For $n_1 \neq n_2 \neq n_3$,

$$\tilde{Q}^{(3)}_{n_1 n_2 n_3}(n) = H_1[z_{n_1}(n)]\, H_1[z_{n_2}(n)]\, H_1[z_{n_3}(n)] \tag{3.83a}$$

For $n_1 = n_2 \neq n_3$,

$$\tilde{Q}^{(3)}_{n_1 n_1 n_3}(n) = H_2[z_{n_1}(n)]\, H_1[z_{n_3}(n)] \tag{3.83b}$$

For $n_1 = n_2 = n_3$,

$$\tilde{Q}^{(3)}_{n_1 n_1 n_1}(n) = H_3[z_{n_1}(n)] \tag{3.83c}$$

It is interesting to note that $\tilde{Q}^{(3)}_{n_1 n_1 n_3}(n) = \tilde{Q}^{(3)}_{n_1 n_3 n_1}(n) = \tilde{Q}^{(3)}_{n_3 n_1 n_1}(n) = H_2[z_{n_1}(n)]H_1[z_{n_3}(n)]$. This implies that $\tilde{Q}$-polynomial does not change for any permutation of components in α, which means that:

$$\tilde{Q}^{(m)}_{\alpha_m}(n) = \tilde{Q}^{(m)}_{\mathrm{perm}(\alpha_m)}(n) \tag{3.84}$$

Because $\tilde{Q}$-polynomial is the nonlinear combination of Hermite polynomials, it also satisfies zero mean and orthogonality properties:

$$E\{\tilde{Q}^{(m)}_{\alpha}(n)\} = 0 \tag{3.85a}$$

$$E\{\tilde{Q}^{(m)}_{\alpha}(n)\tilde{Q}^{(n)}_{\beta}(n)\} = \text{const.}\,\delta(\alpha-\beta)\delta(m-n) \tag{3.85b}$$

Then equation 3.81 can be abbreviated as:

$$G_m[k_m; x(n)] = \sum_{\alpha_m} c_m(\alpha_m)\tilde{Q}^{(m)}_{\alpha_m}(n) \tag{3.86}$$

which means $G_m[k_m;x(n)]$ can be represented by a linear combination of orthogonal $\widetilde{Q}$-polynomials. The sum is from $\alpha_m = \underbrace{0,...,0}_{m's}$ to $\underbrace{n_m,...,n_m}_{m's}$.

The block diagram for $G_m[k_m; x(n)]$ is shown in figure 3-5:

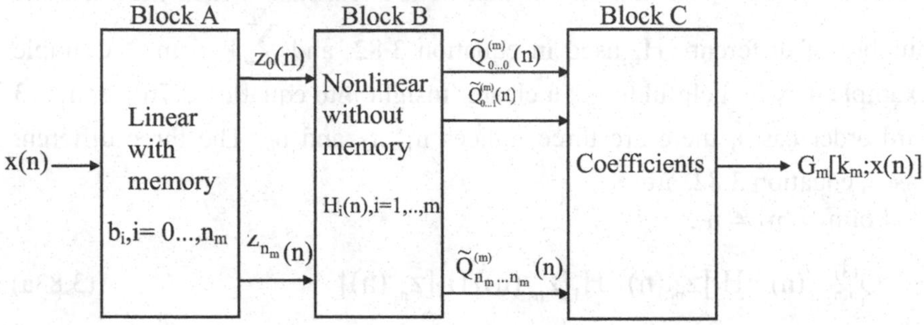

Figure 3-5. Block diagram of $G_m[k_m; x(n)]$

where z = z(n). We note that in block B, starting with $H_0(z) = 1$, the Hermite polynomial has the recursive linear relation:

$$H_{n+1}(z) = zH_n(z) - AH'_n(z) \tag{3.87}$$

where $H'_n(z)$ means the first derivative of $H_n(z)$. The first four polynomials are $H_0(z) = 1$, $H_1(z) = z$, $H_2(z) = z^2\text{-}A$, and $H_3(z) = z^3\text{-}3Az$. Now, substituting equation 3.86 in equation 3.70 yields:

$$y(n) = \sum_{i=0}^{m}\sum_{\alpha_i} c_i(\alpha_i)\widetilde{Q}^{(i)}_{\alpha_i}(n) \tag{3.88}$$

Note that the $\widetilde{Q}$-polynomial satisfies the orthogonal condition. It means that y(n) is a linear combination of a set of orthonormal bases, giving us a great benefit in doing adaptive filtering because the autocorrelation matrix of $\widetilde{Q}$-polynomial is a diagonal matrix.

Consider a special case $G_2[k_2; x(n)]$ with white Gaussian input of σ_x^2 variance; from equation 3.81 we can have:

$$\begin{aligned} G_2[k_2; x(n)] &= c_2(0,0)H_2[z_0(n)] + c_2(0,1)H_1[z_0(n)]H_1[z_1(n)] \\ &\quad + c_2(1,0)H_1[z_1(n)]H_1[z_0(n)] + c_2(1,1)H_2[z_1(n)] \end{aligned} \tag{3.89}$$

From equation 3.82, equation 3.89 can be rewritten as:

$$G_2[k_2; x(n)] = c_2(0,0)\widetilde{Q}_{00}^{(2)}[z(n)] + [c_2(0,1) + c_2(1,0)]\widetilde{Q}_{01}^{(2)}[z(n)] + c_2(1,1)\widetilde{Q}_{11}^{(2)}[z(n)] \quad (3.90)$$

Equation 3.90 can be represented by a block diagram as shown in figure 3-6:

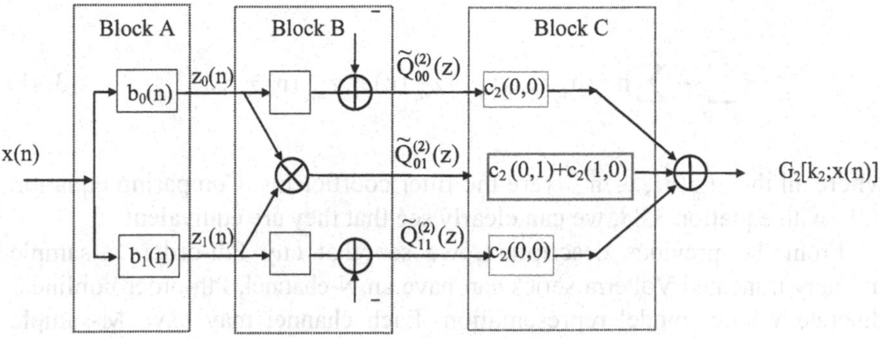

Figure 3-6. Second-order Wiener kernel functional block diagram

3.4 Delay Line Version of Nonlinear Wiener Model

From the above description, we know that all Volterra series can be expanded by using the discrete Wiener model. This means that the Volterra series is a subset of the Wiener class. If a nonlinear system is analytic and can be represented by an Mth-order Volterra series, it must have an equivalent Mth-order Wiener model as shown figure 3-7.

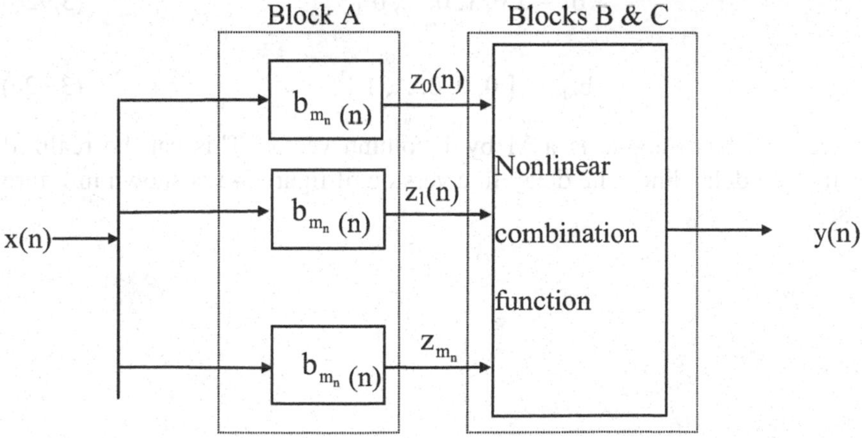

Figure 3-7. Reduced form of the nonlinear Wiener model

Where b_{m_n} are the orthonormal bases. This is the reduced version of figure 3-5. Consequently, we can express the output y(n) by all the nonlinear combination of outputs from orthonormal bases block which is:

$$y(n) = h_0 + \sum_{n_1=0}^{\infty} h_1(n_1) z_{n_1}(n) + \sum_{n_1=0}^{\infty}\sum_{n_2=0}^{\infty} h_2(n_1, n_2) z_{n_1}(n) z_{n_2}(n) + \ldots$$

$$+ \sum_{n_1=0}^{\infty} \cdots \sum_{n_m=0}^{\infty} h_m(n_1, \ldots, n_{m_n}) z_{n_1}(n) \ldots z_{n_m}(n) + \ldots \tag{3.91}$$

where all the $h_m(n_1, \ldots, n_{m_n})$ are the filter coefficients. Comparing equation 3.91 with equation 3.88, we can clearly see that they are equivalent.

From the previous description, we see that any Pth-order M-sample memory truncated Volterra series can have an N-channel, Pth-order nonlinear discrete Wiener model representation. Each channel may have M-sample ($N \leq M$) memory. In the ideal case, if all the orthonormal bases are known, we can express a complicated Volterra series with few coefficients which implies N<M. Practically, these bases may not be known beforehand. We may need to find the full expansion basis set (N = M) in block A. Fortunately, the selection of these bases is quite flexible; we can choose any set of bases if only they are orthonormal. For the Mth-order nonlinear model, the easiest way is to select N = M channels of M-sample basis system in block A which have:

$$b_0 = [1, 0, 0, \ldots, 0]^T \tag{3.92a}$$

$$b_1 = [0, 1, 0, \ldots, 0]^T \tag{3.92b}$$

$$b_{M-1} = [0, 0, 0, \ldots, 1]^T \tag{3.92c}$$

where each basis above is a M by 1 column vector. This can be realized easily by a delay line. The delay line version of figure 3-5 is shown in figure 3-8:

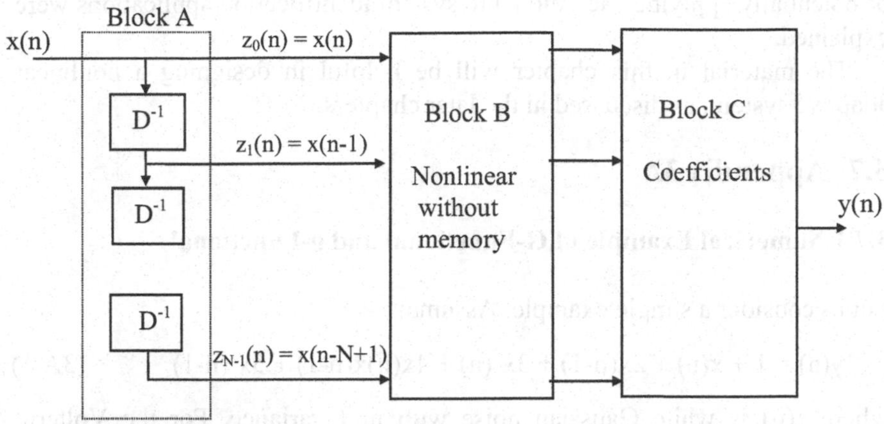

Figure 3-8 Delay line version of nonlinear discrete Wiener model

where D^{-1} means delay input signal by one sample. Even though the bases selection is very flexible, there is no simple solution to get all the nonlinear Wiener model coefficients in figure 3-5 or even in figure 3-8. For the structure in figure 3-8, we can develop a very simple, practical, and efficient LMS-type nonlinear Wiener adaptive algorithm to solve for the coefficients. Detailed derivations, performance analysis, and computer simulations will be illustrated in the next few chapters.

3.5 The Nonlinear Hammerstein Model Representation

There is another representation called the Hammerstein model. In this model, the linear part is sandwiched between two nonlinear parts, unlike the Wiener model, where the nonlinear part is sandwiched between the two linear parts.

We refer the reader to (Rugh WJ 2002), (Westwick K 2003), and others for more details on this model.

3.6 Summary

In this chapter, we have presented the Volterra and Wiener models, both of which are based on truncated Volterra series.

The polynomial modeling of nonlinear systems by these two models was discussed in detail. Relationships between the two models and the limitations of potentially applying each model to system identification applications were explained.

The material in this chapter will be helpful in designing a nonlinear adaptive system, as discussed in the later chapters.

3.7 Appendix 3A

3.7.1 Numerical Example of G-Functional and g-Functional

Let us consider a simple example. Assume:

$$y(n) = 1 + x(n) + 2x(n-1) + 3x^2(n) + 4x(n)x(n-1) + 5x^2(n-1) \quad (3A.1)$$

where x(n) is white Gaussian noise with unit variance. For the Volterra representation in equation 3.12, we have:

$$y(n) = Y_0[x(n)] + Y_1[x(n)] + Y_2[x(n)] \quad (3A\ 2)$$

where $Y_0[x(n)]$, $Y_1[x(n)]$ and $Y_2[x(n)]$ are

$$Y_0[x(n)] = 1 \quad (3A.3a)$$

$$Y_1[x(n)] = [1, 2][x(n), x(n-1)]^T \quad (3A.3b)$$

$$Y_2[x(n)] = [3, 4, 5][x(n)^2, x(n)x(n-1), x^2(n-2)]^T \quad (3A.3c)$$

Note that $Y_0[x(n)]$, $Y_1[x(n)]$ and $Y_2[x(n)]$ are not mutually statistically orthogonal. We can rearrange y(n) by Wiener representation as in equation 3.26:

$$y(n) = g_0[k_0; x(n)] + g_1[k_1, k_{0(1)}; x(n)] + g_2[k_2, k_{1(2)}, k_{2(0)}; x(n)] \quad (3A.4)$$

The $g_0[k_0; x(n)]$, $g_1[k_1, k_{0(1)}; x(n)]$ and $g_2[k_2, k_{1(2)}, k_{2(0)}; x(n)]$ are:

$$g_0[k_0; x(n)] = k_0 \quad (3A.5a)$$

$$g_1[k_1, k_{0(1)}; x(n)] = K_{0(1)}[x(n)] + K_1[x(n)] = k_{0(1)} + k_1[x(n), x(n-1)]^T \quad (3A.5b)$$

$$g_2[k_2, k_{1(2)}, k_{2(0)}; x(n)] = K_{0(2)}[x(n)] + K_{1(2)}[x(n)] + K_{2(2)}[x(n)]$$

$$= k_{0(2)} + k_{1(2)}[x(n), x(n-1)]^T + k_2[x(n)^2, x(n)x(n-1), x^2(n-2)]^T \quad (3A.5c)$$

To satisfy the requirement of equation 3.20, which is:

$E\{Y_0[x(n)]g_1[k_1,k_{0(1)}; x(n)]\} = 0$,

we can find

$k_{0(1)} = 0$,

which implies

$K_{0(1)}[x(n)] = 0$.

From equation 3.36, we know that

$k_{1(2)} = 0$,

which means

$K_{1(2)}[x(n)] = 0$.

Using simple algebra, we can obtain

$k_2 = [3, 4, 5]$.

From equation 3.28a, we require:

$E\{Y_0[x(n)]\, g_2[k_2, k_{1(2)}, k_{2(0)}; x(n)]\} = 0$ and

$E\{Y_1[x(n)]\, g_2[k_2, k_{1(2)}, k_{2(0)}; x(n)]\} = 0$.

This helps us to find that $k_{0(2)} = -8$. Then, we can determine $k_0 = 9$ and $k_1 = [1, 2]$.

Then equation 3A.4 can be rewritten by using the G-functional, which is:

$$y(n) = G_0[k_0;x(n)] + G_1[k_1;x(n)] + G_2[k_2;x(n)] \tag{3A.6}$$

where $G_0[k_0;x(n)]$, $G_1[k_1;x(n)]$ and $G_2[k_2;x(n)]$ are:

$$G_0[k_0;x(n)] = g_0[k_0; x(n)] = 9 \tag{3A.7a}$$

$$G_1[k_1;x(n)] = g_1[k_1; x(n)] = [1, 2][x(n), x(n-1)]^T \tag{3A.7b}$$

$$G_2[k_2;x(n)] = g_0[k_{0(2)}, k_2; x(n)] =$$

$$-8 + [3, 4, 5][x(n)^2, x(n)x(n-1), x^2(n-2)]^T \tag{3A.7c}$$

It can be verified easily by equation 3A.4 that

$G_0[k_0; x(n)]$, $G_1[k_1; x(n)]$ and $G_2[k_2; x(n)]$

are mutually orthogonal.

3.8 Appendix 3B

3.8.1 Kernel k_m of G-Functional

From section 3.2.4, we know that the general form of non-homogeneous mth-order G-functional is:

$G_m[k_m; x(n)] =$

$$\sum_{r=0}^{[m/2]}\sum_{i_1=0}^{\infty}\cdots\sum_{i_{m-2r}=0}^{\infty} k_{m-2r(m)}(i_1, i_2, \ldots, i_{m-2r})\, x(n-i_1)\ldots x(n-i_{m-2r}) \qquad (3B.1)$$

where [m/2] means the largest integer less than or equal to m/2 and

$$k_{m-2r(m)}(i_1, i_2, \ldots, i_{m-2r}) = \frac{(-1)^r\, m!(\sigma_x^2)^r}{(m-2r)!r!2^r}\sum_{j_1=0}^{\infty}\cdots\sum_{j_r=0}^{\infty} k_m(j_1, j_1, \ldots, j_r, j_r, i_1, i_2, \ldots, i_{m-2r}) \qquad (3B.2)$$

Referring to (Schetzen 1980), we can derive equation 3B.2. Recall that $g_m[k_m, \ldots, k_{0(m)}; x(n)]$ is required to be orthogonal to any Volterra functional $Y_n[x(n)]$ whose order is less than m. This implies that the $G_m[k_m; x(n)]$ (canonical form of $g_m[k_m, \ldots, k_{0(m)}; x(n)]$) is also orthogonal to any Volterra functional $Y_i[x(n)]$ whose order is less than m, which means:

$$E\{Y_n[x(n)]G_m[k_m; x(n)]\} = 0 \quad \text{for } n < m \qquad (3B.3)$$

In sections 3.2.1–3.2.3, we see that for an arbitrary Volterra functional kernel, we can always find the kernel in each G-functional which can satisfy the orthogonality requirement. This fact is not only true for first- to third-order cases but also can be extended to the general mth-order case. This means that we can always find the kernel of $G_m[k_m; x(n)]$ to satisfy equation 3B.3 for an arbitrary kernel in each Volterra functional $Y_n[x(n)]$, if each kernel of $Y_n[x(n)]$ is equal to one which can be defined as a delay operator (Schetzen 1980):

$$D_n[x(n)] = x(n-k_1)\, x(n-k_2) \ldots x(n-k_n), \quad \text{for } k_1 \neq k_2 \neq \ldots \neq k_n \qquad (3B.4)$$

Therefore, if we substitute equation 3B.4 in equation 3B.3, equation 3B.3 is still valid:

$$E\{D_n[x(n)]G_m[k_m; x(n)]\} = 0 \qquad \text{for} \quad n < m \tag{3B.5}$$

It is because the delay operator is just a special case of the Volterra model. By using the orthogonality property of equation 3B.5, we can derive equation 3B.2. The procedure is as follows:

In general, $G_m[k_m; x(n)]\}$ can be expressed as:

$$G_m[k_m; x(n)] = \sum_{p=0}^{[m/2]} K_{m-2p(m)}[x(n)] \tag{3B.6}$$

Substituting equation 3B.6 in equation 3B.5, we obtain:

$$E\{D_n[x(n)]G_m[k_m; x(n)]\} =$$

$$\sum_{r=0}^{[m/2]} E\{D_n[x(n)]K_{m-2r(m)}[x(n)]\} = 0, \text{ for } n < m \tag{3B.7}$$

We note that the mean value of the product is zero if n+m-2r is an odd number. This is because the mean value of the product of an odd number of zero-mean Gaussian variables is zero. Therefore, only those for which the value of n+m-2r is an even number need to be considered. Equation 3B.7 can simply be written as (Schetzen 1980):

$$E\{D_n[x(n)]K_{m-2r(m)}[x(n)]\} =$$

$$A[x(n)]\Big|_{\substack{(m-2r)<n \\ n+m-2r=\text{even}}} + B[x(n)]\Big|_{\substack{(m-2r)\geq n \\ n+m-2r=\text{even}}} \tag{3B.8}$$

In equation 3B.8, we note that there are two functionals, A[x(n)] and B[x(n)]. The first functional contains the term m-2r < n. This means that each term in A[x(n)] contains the factor of $E\{x(n-i_j)x(n-i_k)\} = 0$, since $i_j \neq i_k$. Therefore, we know immediately that:

$$A[x(n)]\Big|_{\substack{(m-2r)<n \\ n+m-2r=\text{even}}} = 0 \tag{3B.9}$$

The second functional contains the term m-2r ≥ n. We first pick n i's from m-2r i's and pair them with n k's in the delay operator. Because (n+m-2r) is even, (m-2r-n) is also even. Therefore, we can have q = (m-2r-n)/2 pairs for the rest of (m-2r-n) i's.

There are

$$\binom{m-2r}{2q}\text{'s}$$

ways of picking 2q i's from (m-2r) i's, and there are

$$\frac{(2q)!}{q!2^q}$$

ways of pairing the 2q i's among themselves. Also, there are (m-2r-2q)! ways of pairing each of k's with one of the remaining (m-2r-2q) i's. Therefore, B[x(n)] can be expressed as:

$$B[x(n)]\Big|_{\substack{(m-2r)\geq n\\ n+m-2r=\text{even}}}$$

$$=\binom{m-2r}{2q}\frac{(2q)!}{q!2^q}(m\text{-}2r\text{-}2q)!$$

$$(\sigma_x^2)^{q+n}\sum_{j_1=0}^{\infty}\cdots\sum_{j_q=0}^{\infty}k_{m-2r(m)}(\overbrace{\underbrace{j_1,j_1,\ldots,j_q,j_q}_{\substack{(m-2r-n)'s\\ q=[(m-n)/2]-r}},\underbrace{k_1,k_2,\ldots,k_n}_{n's}}^{(m-2r)'s})$$

$$=\frac{(m-2r)!}{q!2^q}(\sigma_x^2)^{q+n}\sum_{j_1=0}^{\infty}\cdots\sum_{j_q=0}^{\infty}k_{m-2r(m)}(\underbrace{j_1,j_1,\ldots,j_q,j_q}_{(m-2r-n)'s},\underbrace{k_1,k_2,\ldots,k_n}_{n's}) \tag{3B.10}$$

From equation 3B.10, we note that the r range should be from 0 to [(m-n)/2]; then equation 3B.7 can be expressed as:

$$E\{D_n[x(n)]G_m[k_m;x(n)]\}$$

$$=\sum_{r=0}^{[m/2]}E\{D_n[x(n)]K_{m-2r(m)}[x(n)]\}$$

$$=\sum_{r=0}^{[(m-n)/2]}\frac{(m-2r)!}{q!2^q}$$

$$(\sigma_x^2)^{q+n}\sum_{k_1=0}^{\infty}\cdots\sum_{k_n=0}^{\infty}k_{m-2r(m)}(\underbrace{j_1,j_1,\ldots,j_q,j_q}_{(m-2r-n)'s},\underbrace{k_1,k_2,\ldots,k_n}_{n's})=0 \tag{3B.11}$$

To calculate $k_{m-2r(m)}$, first, we consider $n = m-2$ in equation 3B.11, thus:

$$E\{D_{m-2}[x(n)]G_m[k_m; x(n)]\}$$

$$= \frac{m!}{1!2}(\sigma_x^2)^{1+m-2}\sum_{j_1=0}^{\infty} k_m(\underbrace{j_1, j_1}_{2's}, \underbrace{k_1, k_2, \ldots, k_n}_{(m-2)'s}) +$$

$$\frac{(m-2)!}{0!2^0}(\sigma_x^2)^{m-2} k_{m-2(m)}(\underbrace{k_1, k_2, \ldots, k_n}_{(m-2)'s}) = 0 \tag{3B.12}$$

Therefore, $k_{m-2(m)}$ can be obtained:

$$k_{m-2(m)}(k_1, k_2, \ldots, k_{m-2}) = \frac{-m!}{(m-2)!1!2}\sigma_x^2 \sum_{j_1=0}^{\infty} k_m(j_1, j_1, k_1, k_2, \ldots, k_n) \tag{3B.13}$$

Secondly, we consider $n = m-4$ in equation 3B.11, thus:

$$E\{D_{m-4}[x(n)]G_m[k_m; x(n)]\}$$

$$= \frac{m!}{2!2^2}(\sigma_x^2)^{2+m-4}\sum_{j_2=0}^{\infty}\sum_{j_1=0}^{\infty} k_m(\underbrace{j_1, j_1, j_2, j_2}_{4's}, \underbrace{k_1, k_2, \ldots, k_n}_{(m-4)'s})$$

$$+ \frac{(m-2)!}{1!2}(\sigma_x^2)^{1+m-4}\sum_{j_1=0}^{\infty} k_{m-2(m)}(\underbrace{j_1, j_1}_{2's}, \underbrace{k_1, k_2, \ldots, k_n}_{(m-2)'s})$$

$$+ \frac{(m-4)!}{0!2^0}(\sigma_x^2)^{m-4} k_{m-4(m)}(\underbrace{k_1, k_2, \ldots, k_n}_{(m-4)'s}) = 0 \tag{3B.14}$$

Substituting $k_{m-2(m)}$ from equation 3B.13 in equation 3B.14, we obtain $k_{m-4(m)}$ as

$$k_{m-4(m)}(k_1, k_2, \ldots, k_{m-4}) =$$

$$\frac{m!}{(m-4)!2!2^2}(\sigma_x^2)^2 \sum_{j_2=0}^{\infty}\sum_{j_1=0}^{\infty} k_m(j_1, j_1, j_2, j_2, k_1, k_2, \ldots, k_n) \tag{3B.15}$$

Similarly, for n = m-6, we obtain $k_{m-6(m)}$ as:

$$k_{m-6(m)}(k_1,k_2,\ldots,k_{m-6})=\frac{-m!(\sigma_x^2)^3}{(m-6)!3!2^3}$$

$$\sum_{j_1=0}^{\infty}\sum_{j_2=0}^{\infty}\sum_{j_3=0}^{\infty}k_m(j_1,j_1,j_2,j_2,j_3,j_3,k_1,k_2,\ldots,k_n) \quad (3B.16)$$

Accordingly, for n = m-2r, we obtain $k_{m-2r(m)}$ as:

$$k_{m-2r(m)}(k_1,k_2,\ldots,k_{m-2r}) =$$

$$\frac{(-1)^r\, m!\,\sigma_x^{2r}}{(m-2r)!\,r!\,2^r}\sum_{j_1=0}^{\infty}\ldots\sum_{j_r=0}^{\infty}k_m(j_1,j_1,\ldots,j_r,j_r,k_1,k_2,\ldots,k_n) \quad (3B.17)$$

To verify the correctness of equation 3B.17, we need to evaluate whether or not equation 3B.17 is still valid. The procedure is that we substitute equation 3B.17 in equation 3B.10, thus:

$$E\{D_n[x(n)]G_m[k_m;x(n)]\} = \sum_{r=0}^{(m-i)/2}\frac{(m-2r)!}{q!2^q}$$

$$(\sigma_x^2)^{q+n}\sum_{j_1=0}^{\infty}\ldots\sum_{j_q=0}^{\infty}k_{m-2r(m)}(\underbrace{j_1,j_1,\ldots,j_q,j_q}_{(m-2r-n)'s},\underbrace{k_1,k_2,\ldots,k_n}_{n's})$$

$$= \sum_{r=0}^{(m-n)/2}\frac{(m-2r)!}{q!2^q}$$

$$(\sigma_x^2)^{q+n}\frac{(-1)^r\, m!\,\sigma_x^{2r}}{(m-2r)!\,r!\,2^r}\sum_{j_1=0}^{\infty}\ldots\sum_{j_r=0}^{\infty}k_m(\underbrace{j_1,j_1,\ldots,j_r,j_r}_{2r=(m-n)'s},\underbrace{k_1,k_2,\ldots,k_n}_{n's})$$

$$= \sum_{r=0}^{[m-n)/2}\frac{(-1)^r}{([\frac{m-n}{2}]-r)!r!}\frac{m!(\sigma_x^2)^{\frac{m-n}{2}}}{2^{[\frac{m-n}{2}]}}$$

$$\sum_{j_1=0}^{\infty}\ldots\sum_{j_{\frac{m-n}{2}}=0}^{\infty}k_m(j_1,j_1,\ldots,j_{\frac{m-n}{2}},j_{\frac{m-n}{2}},k_1,k_2,\ldots,k_n) \quad (3B.18)$$

We have the expression in equation 3B.18 because r = (m-n)/2. From the binomial expansion, we note that:

$$(1+x)^n|_{x=-1} = \sum_{r=0}^{n}\binom{n}{r}x^r|_{x=-1} = \sum_{r=0}^{n}\frac{n!(-1)^r}{(n-r)!r!} = 0 \qquad (3B.19)$$

Because n! is not related to r, we deduce immediately that:

$$\sum_{r=0}^{n}\frac{(-1)^r}{(n-r)!r!} = 0 \qquad (3B.20)$$

Changing variables by replacing n by (m-n)/2, we see that:

$$\sum_{r=0}^{(m-n)/2}\frac{(-1)^r}{(\frac{m-n}{2}-r)!r!} = 0 \qquad (3B.21)$$

Substituting equation 3B.21 in equation 3B.18, we easily verify that equation 3B.18 is indeed equal to zero.

3.9 Appendix 3C

3.9.1 Separable Property of G-Functional

In this appendix, we show that $G_{k_i+k_j}[b_{m_i}^{(k_i)}b_{m_j}^{(k_j)};x(n)]$ can be separated as:

$$G_{k_i+k_j}[b_{m_i}^{(k_i)}b_{m_j}^{(k_j)};x(n)] = G_{k_i}[b_{m_i}^{(k_i)};x(n)]G_{k_j}[b_{m_j}^{(k_j)};x(n)], \quad \text{for } i \neq j \qquad (3C.1)$$

where $b_{m_i}^{(k_i)}b_{m_j}^{(k_j)}$ denotes the orthonomal bases b_{m_i} and b_{m_j} appear k_i and k_j times respectively. The proof is straightforward.

Proof

From equation 3.69, we know that:

$$G_{k_i}[b_{m_i}^{(k_i)};x(n)] =$$

$$\sum_{m=0}^{[k_i/2]}\frac{(-1)^m k_i!}{m!(k_i-2m)!}\left(\frac{\sigma_x^2}{2}\right)^m\left(b_{m_i}(n)*x(n)\right)^{k_i-2m} \qquad (3C.2)$$

The $G_n[b_{m_i}^{(n)};x(n)]$ is the same as equation 3C.2, except for replacing m_i and $k_i \times k_j$ and m_j. Therefore:

$$G_{k_i}[b_{m_i}^{(k_i)};x(n)]G_{k_j}[b_{m_j}^{(k_j)};x(n)]$$

$$=\sum_{m=0}^{[k_i/2]}\frac{(-1)^m k_i!}{m!(k_i-2m)!}\left(\frac{\sigma_x^2}{2}\right)^m\left(b_{m_i}(n)*x(n)\right)^{k_i-2m}$$

$$\sum_{n=0}^{[k_j/2]}\frac{(-1)^m k_j!}{n!(k_j-2n)!}\left(\frac{\sigma_x^2}{2}\right)^n\left(b_{m_j}(n)*x(n)\right)^{k_j-2n}$$

$$=\sum_{m=0}^{[k_i/2]}\sum_{n=0}^{[k_j/2]}\frac{(-1)^m k_i!}{m!(k_i-2m)!}\frac{(-1)^m k_j!}{n!(k_j-2n)!}\left(\frac{\sigma_x^2}{2}\right)^{m+n}\left(b_{m_i}(n)*x(n)\right)^{k_i-2m}\left(b_{m_j}(n)*x(n)\right)^{k_j-2n}$$

$$=G_{k_i+k_j}[b_{m_i}^{(k_i)}b_{m_j}^{(k_j)};x(n)] \tag{3C.3}$$

The result of equation 3C.3 can easily be extended to the general $b_{m_1}^{(k_1)}b_{m_2}^{(k_2)}\ldots b_{m_M}^{(k_M)}$ case:

For $\sum_{i=1}^{M}k_i = m$, $m_i \neq m_j$ and $k_i \neq k_j$, we have:

$$G_m[b_{m_1}^{(k_1)}b_{m_2}^{(k_2)}\ldots b_{m_M}^{(k_M)};x(n)] =$$

$$G_{k_1}[b_{m_1}^{(k_1)};x(n)]\,G_{k_2}[b_{m_2}^{(k_2)};x(n)]\;\ldots\;G_{k_M}[b_{m_M}^{(k_M)};x(n)] \tag{3C.4}$$

Chapter 4
NONLINEAR SYSTEM IDENTIFICATION METHODS
A brief survey of the available methods

Introduction

In chapter 1, we discussed the importance of nonlinear systems in everyday life. In this chapter, we present the different methods of identifying nonlinear systems based on the model of describing the system. The importance of the model used in describing the nonlinear system is underscored here because the model chosen ultimately determines the quality and type of solution realized. A system identification method is only as good as the model it utilizes.

There are many different types of models for system identification of nonlinear systems. Some of the rich literature includes sources like (Ljung 1999), (Nelles 2000), (Ogunnaike 1994), (Sjoberg 1995), (Diaz 1988), and (Billings 1980). In this book we can only touch on some of the methods that are recent and relevant to us.

First we summarize other methods, both those based on nonlinear local optimization and those based on nonlinear global optimization. Then we present one of the popular recent neural network approaches based on the Wiener model used in parallel with a neural network (Ibnkahla 2003). Finally, we discuss the need for adaptive methods and our methods.

4.1 Methods Based on Nonlinear Local Optimization

A nonlinear function of the parameter (weight) vector can be searched by using nonlinear local optimization techniques. Some of the properties of the nonlinear optimization problem are: (1) there are many local optimal points (minima), and (2) the surface around a local optimum can usually be approximated by a second-order Taylor series expansion. This second property helps us use the Taylor series (and related Volterra series) expansions for finding the optimum points.

The methods discussed in this section start at an initial point and search for a local optimum around the neighborhood of the initial point. To perform a global optimization search, we will need to start a local search from many different starting points and choose the best solution.

Therefore it is very important to pick a very good starting point for our local search.

Examples of possible search methods based on nonlinear local optimization are discussed as follows:

(1) Direct search (examples of which are simplex search method and Hooke-Jeeves method). These methods do not require that the derivatives of the cost function (sometimes also called the *loss function*) are available. Instead the search is based on computing the cost function at each point in the trajectory until the optimal point is found. The algorithms usually converge very slowly.

(2) General gradient-based algorithms (examples of which are line search, finite-difference techniques, steepest-descent, Newton's method, Quasi-Newton methods, and conjugate-gradient methods). These methods are the most common and effective methods for nonlinear local optimization.

The gradient vector of the cost function (*loss function*) $J(\theta)$ with respect to the parameter vector θ is:

$$g = \frac{\partial J(\theta)}{\partial \theta}$$

This gradient is known or can be approximated. The parameter vector is updated at time n by:

$$\theta_n = \theta_{n-1} - \mu_{n-1} p_{n-1} \quad \text{where } p_{n-1} = R_{n-1} g_{n-1}$$

where μ_{n-1} is a variable step size which controls the rate of convergence to the optimal solution. Here the new parameter vector is chosen by modifying the old parameter vector by a weighted amount in the negative direction of the gradient vector g_{n-1}. The weighting on the gradient vector R_{n-1} is chosen as a positive definite direction matrix which guarantees that $J(\theta_n) < J(\theta_{n-1})$.

A choice of $R_{n-1} = I$ leads to the steepest descent algorithm, a precursor to the least mean square (LMS) algorithm discussed in chapter 5. There are several other possible choices for R_{n-1}.

Newton's method requires a choice of the direction of the matrix R_{n-1} as the inverse of the Hessian matrix ($R_{n-1} = H_{n-1}^{-1}$) of the cost function at the point θ_{n-1}. To get the Hessian requires that second-order derivatives are existent.

Quasi-Newton methods approximate the Hessian matrix in Newton's method. The parameter vector is updated at time n by:

$$\theta_n = \theta_{n-1} - \mu_{n-1} p_{n-1} \text{ where } p_{n-1} = \hat{H}_{n-1} g_{n-1}$$

where the Hessian can be updated using either

$$\hat{H}_n = \hat{H}_{n-1} + Q_{n-1} \quad \text{or} \quad \hat{H}_n^{-1} = \hat{H}_{n-1}^{-1} + \tilde{Q}_{n-1}.$$

These types of algorithms are characterized by (1) fast convergence, (2) no need for second-order derivatives for the Hessian, (3) low computational complexity, and (4) high memory requirements, especially for large problems.

Conjugate gradient methods are rough approximations of the Quasi-Newton method where there is no direct approximation to the Hessian. Instead, the parameter vector is updated by:

$$\theta_n = \theta_{n-1} - \mu_{n-1} p_{n-1} \text{ where } p_{n-1} = g_{n-1} - \beta_{n-1} p_{n-2}$$

where the scalar factor β_{n-1} can be determined in a few different ways. This leads to different variations of this algorithm. One of the most popular variations is the Fletcher-Reeves method (Gill 1981, Nelles 2000).

(3) Nonlinear least squares methods (examples of which are the Gauss-Newton method and the Levenberg-Marquardt method). These methods minimize the cost function

$$J(\theta) = \sum_{i=1}^{n} \lambda^{n-i} \,|\, f(i,\theta) \,|^2$$

where λ^{n-i} is an exponential weighting factor.

If the function $f(i,\theta)$ is a linear function of parameter θ, the linear least squares problem results (see chapter 5). But if the function $f(i,\theta)$ is a *nonlinear* function of parameter θ, then we have the nonlinear least squares problem.

When the linear least squares method is used, the following are the advantages: (1) a unique global optimum point can be found, (2) the estimation variance can be computed easily, (3) recursive formulation can be used for the least squares solution, and (4) ridge selection and subset selection can be done for the trade-off in reduction of bias and variance. When the nonlinear least squares method is used instead, these useful properties may not be retained.

Depending on the overall structure used, we can have $f(i, \theta) = e(i) = d(i) - y(i)$, , which is the error between the desired input and the estimated value.

The gradient and the Hessian matrix of this cost function $J(\theta)$ are used in developing the algorithms based on the Gauss-Newton method and the Levenberg-Marquardt method. The Gauss-Newton method is a nonlinear counterpart of the Newton method discussed above where

$$\theta_n = \theta_{n-1} - \mu_{n-1} p_{n-1} \quad \text{where} \quad p_{n-1} = H_{n-1}^{-1} g_{n-1}$$

Here, the Hessian is approximated by the product of the Jacobian ϑ_{n-1} matrices $H_{n-1}^{-1} \approx (\vartheta_{n-1}^T \vartheta_{n-1})^{-1}$ and the gradient represented by $g_{n-1} = \vartheta_{n-1}^T f(\theta_{n-1})$.

Chapter 10 deals with the application of the least squares method (RLS algorithm) to the Wiener nonlinear model for adaptive filtering.

(4) Constrained nonlinear optimization methods. Here the task is to minimize the cost function subject to constraints. For example, we wish to minimize the cost function $J(\theta)$ with respect to the parameter vector θ subject to constraints:

$$g_i(\theta) \le 0 \quad i = 1, 2, \ldots\ldots, m$$
$$h_j(\theta) = 0 \quad j = 1, 2, \ldots\ldots, l$$

The Lagrangian cost function to be minimized is now:

$$L(\theta, \lambda) = J(\theta) + \sum_{i=1}^{m} \lambda_i g_i(\theta) + \sum_{j=1}^{l} \lambda_{j+m} h_j(\theta)$$

The necessary and sufficient conditions for optimality of this cost function can be given by the famous Kuhn-Tucker equations (Gill 1981, Nelles 2000).

In summary, the direct search, steepest descent, Newton, regularized Newton, quasi-Newton, and conjugate gradient methods are all based on minimizing the general cost functions for nonlinear local optimization methods. But the Gauss-Newton and Levenberg-Marquardt methods are nonlinear least-squares methods.

4.2 Methods Based on Nonlinear Global Optimization

As mentioned in the previous section, a nonlinear function of the parameter (weight) vector can be searched by using nonlinear local optimization techniques. Linear optimization problems typically have a unique global optimum. However, many times, the nonlinear function of the parameter (weight) vector has multiple minima. In addition, it may not be possible to approximate the surface around a local optimum by a second-order Taylor series expansion. The derivatives of the cost function may be noncontinuous.

The methods discussed in this section start at an initial point and search for a global optimum. For some of the methods, the search will not produce a globally optimum point but a locally optimum one, especially for problems with many parameters. Instead we aim to find the best local optimum point out of the many. So we use a multi-start approach where we run each of the local optimization algorithms with different initial conditions. We pick the best result as the global optimum. A truly global search may be very expensive computationally, and sometimes impossible, if we do not have a cost function value at the global optimum point.

Many global search methods mentioned here examine the whole parameter space, thereby requiring lots of computational power, and they converge very slowly. Some also introduce the use of stochastic variables in order to allow the algorithm to escape easily from local optima.

Examples of nonlinear global optimization algorithms are (1) simulated annealing; (2) evolutionary algorithms (EA), which include the methods of evolutionary strategies, genetic algorithms, and genetic programming; (3) branch and bound (B&B) methods; and (4) Tabu search. For more details of these algorithms see (Nelles 2000).

For best results, it is common practice to combine global search methods with nonlinear local or linear optimization methods, usually by incorporating any prior knowledge into the algorithm.

4.3 Neural Network Approaches

Recently, neural network approaches have been used for identification of memory-less nonlinear systems. Some results have shown good performance (Ibnkahla 2002, 2003).

One of the recent neural network approaches is shown in figure 4-1. It is based on the Wiener model used in parallel with a neural network. The Wiener model was presented in chapters 2 and 3.

The unknown nonlinear system is comprised of a nonlinear Wiener system with a linear filter, H(.), followed by a memory-less nonlinearity, g(.). The structure identifies the linear filter H by the adaptive filter Q. And it models the nonlinearity g(.) by using the memory-less nonlinear neural network, which is a two-layer adaptive neural network. There are *M* neurons in the input layer. Each neuron may be represented by an exponential or a saturation nonlinearity (Haykin 1998).

The nonlinear system output signal is corrupted by a zero-mean additive white Gaussian noise N_0 (n); it can be expressed at time *n* as:

$$d(n) = g\left(\sum_{k=0}^{N_h-1} h_k x(n-k)\right) + N_0(n)$$

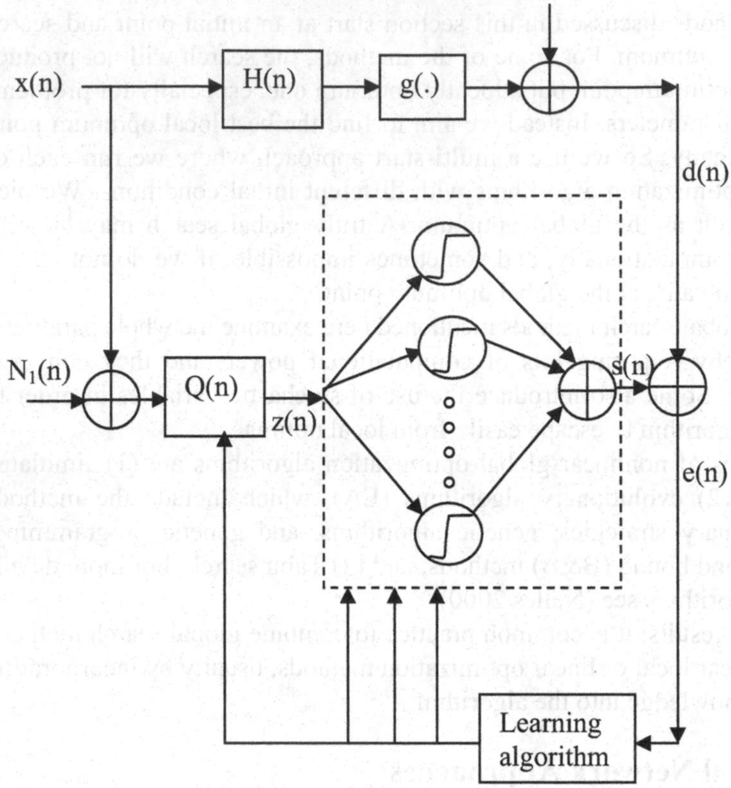

Figure 4-1. A structure for system identification using neural networks

The FIR adaptive filter can be represented by the Z transform:

$$Q(z) = \sum_{k=0}^{N_q-1} q_k z^{-k}$$

The system has a scalar (real-valued) input and a scalar output.

The overall system is considered an unknown nonlinear system, or a "black box." This means the learning is performed using the input-output signals only.

The learning algorithm used can be one of the following methods:

1. LMS-back-propagation algorithm;
2. natural gradient learning;
3. separately updating the linear filter coefficients and the neural network weights using natural gradient learning and LMS back-propagation algorithms respectively or together depending on the definition of the parameter space.

The details of the LMS algorithm are discussed in chapter 5.

There are other possible combinations of algorithms to use for adapting this structure. Some of them include:

1. using the classical LMS algorithm for the adaptive filter and the back-propagation algorithm for the neural network;
2. using the classical LMS algorithm for the adaptive filter and natural gradient algorithm for the neural network;
3. using the natural gradient LMS algorithm for the adaptive filter and back-propagation algorithm for the neural network;
4. using the natural gradient LMS algorithm for the adaptive filter and natural gradient algorithm for the neural network (here we consider the parameter space as a single space—this is known as the coupled NGLMS-NGBP algorithm);
5. using the natural gradient LMS algorithm for the adaptive filter and natural gradient algorithm for the neural network (here both parameter spaces are separated—this is known as the disconnected NGLMS-NGBP algorithm.

It was shown in (Ibnkahla 2003) that the algorithms in combination 5 performed the best among these choices.

An example of such a nonlinear system is a nonlinear channel which corresponds to a typical uplink satellite communication channel. It is composed of a linear filter followed by a traveling wave tube (TWT) amplifier (Ibnkahla 2003) where, for example, the nonlinearity was

$$g(x) = \frac{\alpha x}{1+\beta x^2}, \quad \alpha = 2, \beta = 1.$$

where the input signal can be assumed to be white Gaussian with a variance of 1. Filter H(.) weights are taken as

$$H = [1.4961, 1.3619, 1.0659, 0.6750, 0.2738, -0.0575,$$
$$-0.2636, -0.3257, -0.2634, -0.125]^T$$

Typical values were chosen for the other parameters.

The same structure can be found in other applications: for example, in microwave amplifier design when modeling solid state power amplifiers (SSPA), in adaptive control of nonlinear systems, and in biomedical applications when modeling the relationships between cardiovascular signals.

The need for adaptive methods

The methods described in the previous sections can be useful in certain cases. However, as we have described, all these methods have some limitations.

As discussed in chapter 3, there is a special relationship between the Volterra series model and the Wiener series model. This means that the classical methods used for system identification based on the Volterra series model can also be used for the Wiener series model. In addition, these methods can be made adaptive for cases where the underlying system, system structure, or signal environment is changing. We will explore these methods in later chapters.

4.4 Summary

In this chapter, we have summarized other methods based on nonlinear local optimization as well as those based on nonlinear global optimization. Then we discussed one of the popular recent neural network approaches based on the Wiener model used in parallel with a neural network. Finally, we discussed the need for adaptive methods.

In the next chapter, we present an introduction to adaptive signal processing methods which can be used in nonlinear adaptive system identification.

Chapter 5

INTRODUCTION TO ADAPTIVE SIGNAL PROCESSING

Introduction

In this book, the focus is on adaptive system identification methods for nonlinear systems. But first we need to review some important details about linear adaptive filtering (or adaptive signal processing).

Adaptive filters have been popular since the early 1960s after they were studied and developed by (Widrow 1959). His development is based on the theory of Wiener filters for optimum linear estimation. There are other approaches to the development of adaptive filter algorithms, such as Kalman filters, least squares, etc. We will discuss the theory of Wiener filters and least squares as they relate to adaptive filters. The subject of Kalman filters is beyond the scope of this book. However, it has recently been shown that there is a close relationship between Kalman filters and recursive least squares adaptive filters (Sayed K 1994, Sayed A 2003, Diniz 2002).

In this chapter, we first develop the classical Wiener filter. Next we develop the stochastic gradient adaptive filter in the form of the least-mean square (LMS) algorithm. There are many variants of the LMS algorithm. Then we present the least-squares estimation method and the recursive least-squares (RLS) adaptive algorithm.

5.1 Wiener Filters for Optimum Linear Estimation

The work of Norbert Wiener spans the area of linear and nonlinear filters. For linear filters, he developed the well known Wiener filter upon which the theory of adaptive filters emerged. Also as we showed in chapter 2 and later on in chapter 7, his Wiener model for memory-less nonlinear systems has been shown to realize truncated Volterra series models for nonlinear systems.

The classical Wiener filter is used for optimum linear estimation of a desired signal sequence from a related input signal sequence, as shown in figure 5-1

The filter in figure 5-1 can be a finite impulse response (FIR) filter or an infinite impulse response (IIR) filter. We will assume a FIR filter for this derivation.

We assume the knowledge of certain statistical parameters (e.g., mean and correlation functions) of the input signal and unwanted additive noise. The Wiener filter solves the problem of designing a linear filter with the noisy data as input and the requirement of minimizing the effect of the noise at the filter output according to some statistical criterion.

An error signal is generated by subtracting the desired signal from the output signal of the linear filter.

The statistical criterion often used is the mean-square-error (MSE).

$$\xi = E[| e(n) |^2] \tag{5.1}$$

A desirable choice of this performance criterion (or cost function) must lead to tractable mathematics and to a single optimum solution. For the choice of the MSE criterion and FIR filter, we get a performance function that is quadratic, which means the optimum point is single (see figure 5-2). For the choice of the MSE and IIR filter, we typically get a nonquadratic performance function which is multi-modal, having many minima.

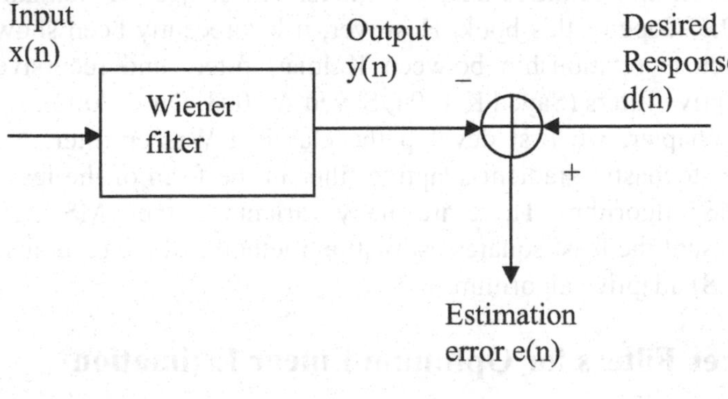

Figure 5-1. The Wiener filter

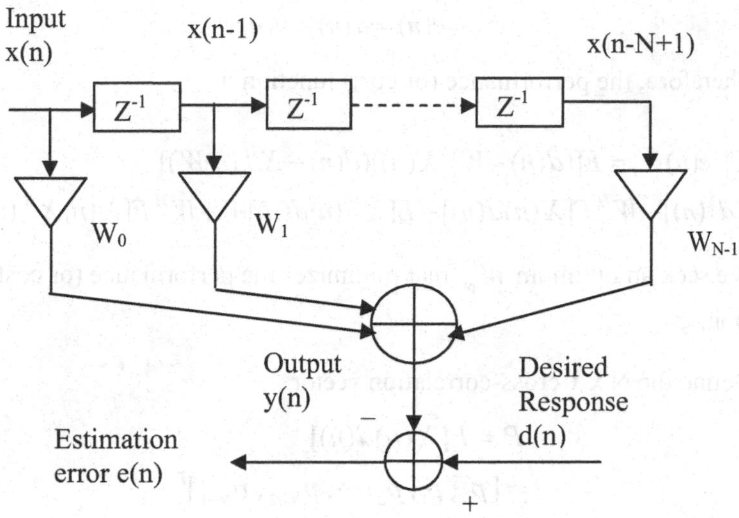

Figure 5-2. Wiener filter based on the FIR filter structure

In figure 5-2,

$$W(n) = [w_0(n), w_1(n), w_2(n), \cdots, w_{N-2}(n), w_{N-1}(n)]^T$$

When we assume a fixed weight vector, we can define

$$W = [w_0, w_1, w_2, \cdots, w_{N-2}, w_{N-1}]^T .$$

Let

$$X(n) = [x_0(n), x_1(n), x_2(n), \cdots, x_{N-2}(n), x_{N-1}(n)]^T ,$$

which is a set of N inputs.

Alternatively for an FIR filter, we define the input vector as a set of delayed N inputs.

$$X(n) = [x(n), x(n-1), x(n-2), x(n-3), \cdots, x(n-N+2), x(n-N+1)]^T$$

The output is

$$y(n) = W^T(n)X(n) \tag{5.2}$$

The error is

$$e(n) = d(n) - y(n) \tag{5.3}$$

Therefore, the performance (or cost) function is

$$\xi = E[| e(n) |^2] = E[(d(n) - W^H X(n))(d(n) - X^H (n)W)]$$
$$= E[(d^2 (n)] - W^H E[X(n)d(n)] - E[X^H (n)d(n)]W + W^H E[X(n)X^H (n)]W)]$$

We seek an optimum W_{opt} that minimizes the performance (or cost) function, ξ.

Define the N x 1 cross-correlation vector

$$P = E[X(n)d(n)] \tag{5.4}$$
$$= [p_0, p_1, p_2, \cdots, p_{N-2}, p_{N-1}]^T$$

The N x N auto-correlation matrix

$$R = E[X(n)X^H (n)] \tag{5.5a}$$
$$= \begin{bmatrix} r_{00} & r_{01} & \cdots & r_{0,N-1} \\ r_{10} & r_{11} & \cdots & r_{1,N-1} \\ \vdots & \vdots & \ddots & \vdots \\ r_{N-1,0} & r_{N-1,1} & \cdots & r_{N-1,N-1} \end{bmatrix}$$

where $r_{i,j}$ is the auto-correlation function given by

$$r_{i,j} = E[x(n-i)(n-j)].$$

For a lag of k, $r(k) = E[x(n)x^*(n-k)]$ and $r^*(k) = E[x^*(n)x(n-k)]$ are the auto-correlation functions. Therefore,

$$R = E[X(n)X^H (n)] \tag{5.5b}$$
$$= \begin{bmatrix} r(0) & r(1) & \cdots & r(N-1) \\ r^*(1) & r(0) & \cdots & r(N-2) \\ \vdots & \vdots & \ddots & \vdots \\ r^*(N-1) & r^*(N-2) & \cdots & r(0) \end{bmatrix}$$

The matrix R has some very interesting properties.

Properties of the auto-correlation matrix R

1. If the input x(n) is a stationary, discrete-time stochastic process, then R is Hermitian. This means $R^H = R$.
   For a stationary input, the auto-correlation function
   $r(k) = r^*(-k)$ where $r(k) = E[x(n)x^*(n-k)]$ and
   $r(-k) = E[x(n)x^*(n+k)]$.

2. If the input x(n) is a stationary, discrete-time stochastic process, then R is Toeplitz. This means the left diagonal elements are the same along any diagonal.

3. If the input x(n) is a stationary, discrete-time stochastic process, then R is non-negative definite and most likely will be positive definite. This means all the eigenvalues of the R matrix are greater than or equal to zero.

4. If the input x(n) is a wide-sense stationary, discrete-time stochastic process, then R is non-singular.

5. If the input x(n) is a stationary, discrete-time stochastic process, then $E[X^R(n)X^{RH}(n)] = R^T$ where
   $X(n) = [x(n), x(n-1), x(n-2), x(n-3), \cdots, x(n-N+2), x(n-N+1)]^T$
   is the input data vector and
   $X^R(n) = [x(n-N+1), x(n-N+2), \ldots\ldots, x(n-1), x(n)]^T$ is the reversed, transposed input data vector.

$$R^T = E[X^R(n)X^{RH}(n)]$$
$$= \begin{bmatrix} r(0) & r^*(1) & \cdots & r^*(N-1) \\ r(1) & r(0) & \cdots & r^*(N-2) \\ \vdots & \vdots & \ddots & \vdots \\ r(N-1) & r(N-2) & \cdots & r(0) \end{bmatrix}$$

   This is a result of the stationary property of the input. Note also that because the R matrix is Hermitian, then the R matrix of the reversed input vector $X^R(n)$ is the same as that of the original input vector $X(n)$.

6. If the input x(n) is a stationary, discrete-time stochastic process, then R can be partitioned as follows:

$$R_{N+1} = \begin{bmatrix} r(0) & r^H \\ r & R_N \end{bmatrix}$$

or as

$$R_{N+1} = \begin{bmatrix} R_N & r^{R*} \\ r^{RT} & r(0) \end{bmatrix}$$

r(0) is the auto-correlation function of the input x(n), for zero lag.

$$\begin{aligned} r^H &= E[x(n)X^H(n)] \\ &= [r(0), r(1), \cdots, r(N-2), r(N-1)] \end{aligned}$$

and

$$\begin{aligned} r^{RT} &= E[x(n)X^{RH}(n)] \\ &= [r(-N), r(-N+1), \cdots, r(-2), r(-1)] \end{aligned}$$

There are other properties but these are the ones which are important for us for our purposes here.
Now note that

$$E[d(n)X^H(n)] = P^H$$
$$W^H P = P^H W$$

Recall that for a linear prediction application (discussed later in section 5.3), $d(n) = x(n)$. Then the N x 1 cross-correlation vector for a FIR filter is

$$\begin{aligned} P &= E[X(n)d(n)] = E[X(n)x(n)] \\ &= [r(0), r(1), \cdots, r(N-2), r(N-1)]^T \end{aligned}$$

Then we obtain

$$\xi = E[(d^2(n)] - 2W^H P + W^H RW \qquad (5.6)$$

This is a quadratic function of the weight vector. Therefore the optimum solution will be a single location in the performance function space corresponding to the optimum weight vector we are seeking.

To minimize this quadratic function, we differentiate it with respect to the weight vector and set the result to the zero vector

$$\nabla_W \xi = \frac{\partial \xi}{\partial W} = 0, \quad \text{for } i = 0, 1, \ldots.., N-1$$

where the gradient vector is a column vector defined as

$$\nabla_W = [\frac{\partial}{\partial W_0} \quad \frac{\partial}{\partial W_1} \quad \cdots\cdots \quad \frac{\partial}{\partial W_{N-1}}]^T$$

Using vector differentiation (Sayed A 2003, Haykin 1996, Diniz 2002), we get

$$\frac{\partial \xi}{\partial W} = -2P + 2RW \tag{5.7}$$

Setting this to zero and solving for W_{opt}, we get

$$\boxed{W_{opt} = R^{-1}P} \tag{5.8}$$

This is the Wiener-Hopf equation that minimizes the mean-square-error (MSE) performance function.

The minimum value of the performance function, ξ, is

$$\xi_{min} = E[d^2(n)] - W_{opt}{}^T P = E[d^2(n)] - W_{opt}{}^T R W_{opt} \tag{5.9a}$$

Also,

$$\xi_{min} = E[d^2(n)] - W_{opt}{}^T P = E[d^2(n)] - P^T R^{-1} P \tag{5.9b}$$

The principle of orthogonality applies to the Wiener filter. At the optimum point, the estimation error, $e_o(n)$, is uncorrelated with the filter tap inputs $x(n-i)$. This means

$$E[e_0(n)x(n-i)] = 0, \quad \text{for } i = 0,1,....,N-1$$

A corollary to this is also that the filter output is orthogonal to the estimation error at the optimum point:

$$E[e_0(n)y_0(n)] = 0$$

Also, note that

$$\xi_{min} = E[d^2(n)] - E[y_0{}^2(n)]. \tag{5.10}$$

Implementing this Wiener-Hopf equation is difficult in practice because it requires knowledge of the auto-correlation matrix of the input vector and the cross-correlation matrix of the input vector and the desired response. Many times the statistical properties of the input and the desired response are not known.

Also, the computational requirement of inverting the R matrix is quite demanding. For a matrix of size NxN, it is $O(N^3)$.

Therefore, there is a need for a recursive determination of the optimum weight vector that gives us the minimum of the quadratic performance function. This leads to the necessity for adaptive filters. This is discussed in the next section.

5.2 Adaptive Filters (LMS-Based Algorithms)

Adaptive filters are useful for situations where the characteristics of the input signals or the system parameters are unknown or time-varying. The design of the filter has to continually *adapt* to the changing environment.

Figure 5-3 illustrates the requirements of an adaptive filter, namely: a filter structure, a filter performance measure and an adaptive update algorithm.

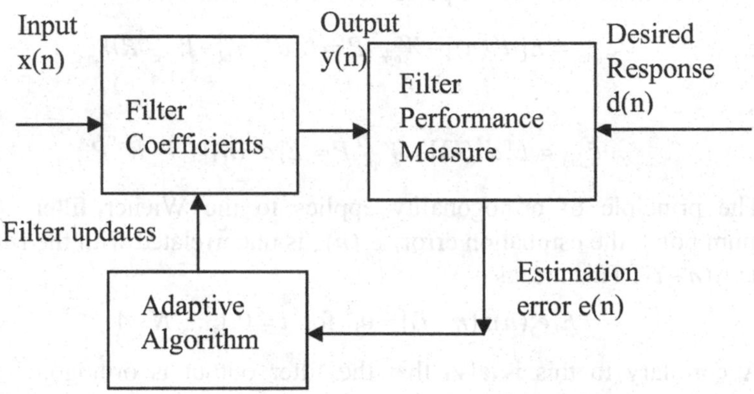

Figure 5-3. A generic adaptive filter

where the input signal is x(n), the desired response is d(n), the error signal is e(n) and the output signal is y(n).

The output of the adaptive filter y(n) is compared with the desired response d(n). The error signal generated is used to adapt the filter parameters to make y(n) more closely approach d(n) or equivalently to make the error e(n) approach zero.

The minimization of a function of the error signal is used for the design. The choice of the performance function depends on the application. There are a variety of possible functions, but the most popular is the mean-square-error (MSE) function because it leads to a quadratic error surface. Recall the MSE is

$$\xi(n) = E[|e(n)|^2] \tag{5.1}$$

We also have a choice of the filter structure. Examples of possible filter structures are FIR (finite impulse response), IIR (infinite impulse response), systolic array, lattice, etc.

If we choose FIR, then we can write

$$y(n) = WX^T(n) \tag{5.2}$$

$$e(n) = d(n) - y(n) \tag{5.3}$$

We substitute equation 5.2 and 5.3 in 5.1. We minimize 5.1 just as in section 5.1. The steepest gradient descent method of minimization requires that we update the weight vector in the negative direction of the steepest gradient, according to the following formula:

$$\boxed{W(n+1) = W(n) - \mu \frac{\partial \xi}{\partial W}} \tag{5.11}$$

This means we change the weight vector in the direction of the negative gradient at each location on the performance surface at each instance of time, n. The amount of change is proportional to the unit-less constant, μ, also sometimes called the step-size. Sometimes it is time-varying $\mu(n)$.

For the FIR filter with a stationary input, the gradient of the cost function is as shown earlier:

$$\frac{\partial \xi}{\partial W} = -2P + 2RW$$

Substituting this in the above formula, the resulting algorithm is

$$\boxed{W(n+1) = W(n) + 2\mu(P - RW)} \tag{5.12}$$

This is the steepest-descent adaptive algorithm. However, this also requires the knowledge of R and P, which are statistical quantities, difficult to determine in practice.

The least-mean-square (LMS) algorithm can be derived by either of two approaches:

(1) Substituting the instantaneous performance (or cost) function $\xi(n) = J(n) = |e(n)|^2$ for the original one, $\xi = E[|e(n)|^2]$.

(2) Using a gradient estimate instead of the actual gradient derived above. This will mean using estimates to compute R and P in $\partial\xi/\partial W = -2P + 2RW$.

Both approaches lead to the same result: the LMS algorithm.

Using the first method:

$$\hat{\xi}(n) = J(n) = |e(n)|^2 = e(n)e^*(n)$$

$$\frac{\partial J}{\partial W} = e(n)\frac{\partial e^*(n)}{\partial W} + e^*(n)\frac{\partial e(n)}{\partial W}$$

$$\frac{\partial e^*(n)}{\partial W} = -X^H(n), \frac{\partial e(n)}{\partial W} = -X(n)$$

Therefore, $W(n+1) = W(n) - \mu\frac{\partial J}{\partial W}$ resulting in

$$\boxed{W(n+1) = W(n) + 2\mu e^*(n)X(n)} \tag{5.13}$$

Using the second method:
An estimate of the true gradient is

$$\hat{\nabla}_W = \partial\hat{\xi}\big/\partial W = -2(\hat{P} - \hat{R}W)$$

where the new parameters used are instantaneous estimates of R and P:

$$\hat{P} = X(n)d^*(n)$$

and

$$\hat{R} = X(n)X^H(n)$$

Therefore,

$$W(n+1) = W(n) - \mu\frac{\partial\xi}{\partial W} = W(n) - \mu(\hat{\nabla}) = W(n) + 2\mu(\hat{P} - \hat{R}W)$$

$$\begin{aligned} W(n+1) &= W(n) + 2\mu(X(n)d^*(n) - X(n)X^H(n)W) \\ &= W(n) + 2\mu X(n)[d^*(n) - X^H(n)W] \end{aligned} \tag{5.14}$$

Again, the resulting weight-update algorithm is

$$\boxed{W(n+1) = W(n) + 2\mu e^*(n)X(n)} \tag{5.15}$$

Another example of an alternative performance (or cost) function is

$$\xi(n) = E[|e(n)|^{2K}] \text{ where } K = 2,3,4,\ldots\ldots \tag{5.16}$$

When K=2, we get the least-mean-fourth (LMF) algorithm.

For more details of the LMF algorithms, see the references (Walach 1984, Chang 1998a).

There are of course other examples of performance functions, many of which do not lead to quadratic performance surfaces.

Convergence of the LMS adaptive filter can typically be analyzed in the mean or in the mean-square. More details of convergence analysis can be found in (Diniz 2002, Haykin 1996, and Sayed A 2003).

5.3 Applications of Adaptive Filters

Adaptive filtering techniques can be applied in four different ways:

(1) system identification,
(2) linear prediction,
(3) channel equalization (inverse filtering/modeling),
(4) interference (noise) cancellation.

The first application is for system identification, as shown in figure 5-4. Here the adaptive filter's desired response is the output of the unknown system. The other ways of applying adaptive filters differ in the way the desired response is determined.

Figure 5-5 shows the application of the adaptive filter for prediction. Here the desired response d(n) is the same as the input x(n), but the input to the adaptive filter is a delayed version of the input x(n). At the convergence of the adaptive filter, the adaptive filter will have been able to correctly predict the actual input using the delayed input.

Figure 5-6 shows the application of the adaptive filter for inverse modeling which is used for channel equalization. Here the adaptive filter is in series with the channel and is expected to provide an inverse to the channel such that the output y(n) is only a delayed version of the original input x(n). Also, here the desired response d(n) is formed by delaying x(n).

The last application of adaptive filters is in interference (noise) cancellation. Here in figure 5-7, there are two inputs to the adaptive filter.

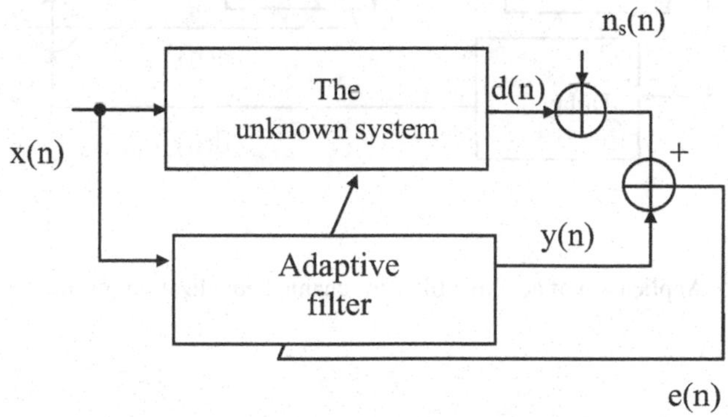

Figure 5-4. Application of adaptive filter for system identification

The two inputs are correlated. The adaptive filter acts on one input to de-correlate it from the other. If one input is a noisy version of the other, then this will result in interference cancellation.

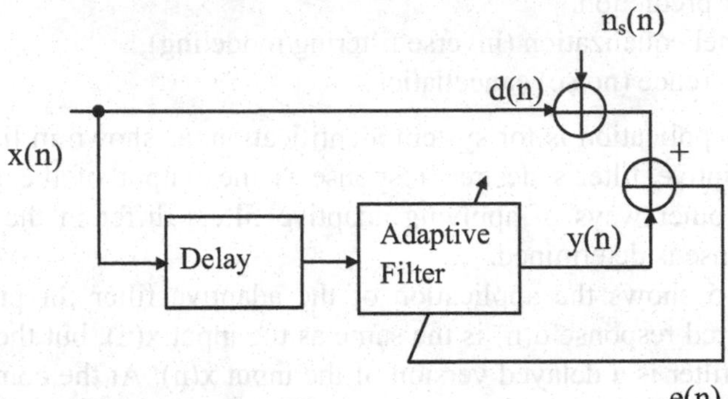

Figure 5-5. Application of adaptive filter for prediction

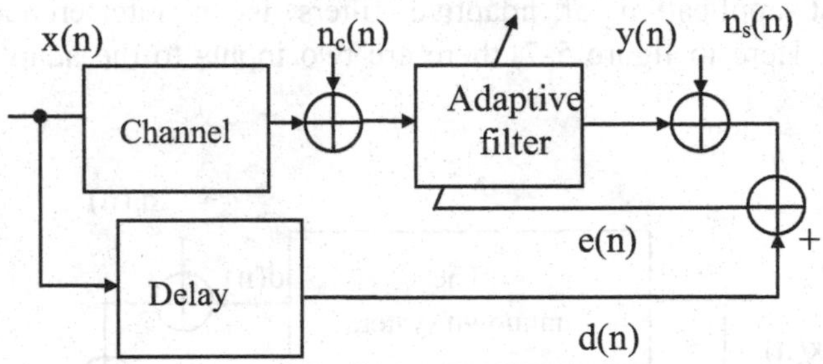

Figure 5-6. Application of adaptive filter for channel equalization (or inverse modeling)

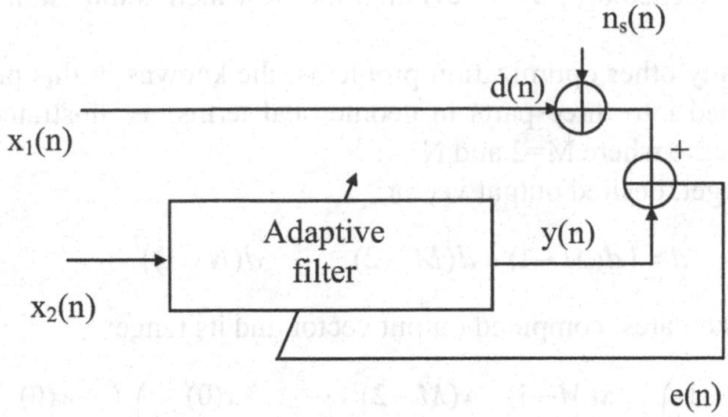

Figure 5-7. Application of adaptive filter for interference cancellation

5.4 Least-Squares Method for Optimum Linear Estimation

If we are given a desired sequence d(n) and an input data sequence x(n), n=0,1,…, N-1 where N is the total number of samples. It is assumed that d(n) and x(n) are samples of stochastic variables.

The problem is to find a filter coefficient (vector) sequence w(k), k=0,1,…, M-1 such that when the input data sequence is filtered, the output best matches the desired sequence, in the sense that the sum of the squared errors will be minimal.

$$J(w) = \sum_{n=M-1}^{n=N-1} |e(n)|^2$$

where

$$e(n) = d(n) - y(n) \tag{5.17}$$

$$y(n) = \sum_{k=0}^{k=M-1} w_k x(n-k)$$

In this problem statement, we need to specify the unknowns that are to be found and the knowns that are needed for computing the unknowns. In this case, we have an *optimal filtering* problem. The unknowns are filter coefficients and the knowns include filter input data and a performance metric for the filter output. Specifically, the filter output is measured against a desired sequence d[n] by a deviation metric which is the "sum of squared errors."

Like many other optimization problems, the knowns in this problem can be partitioned into three parts in geometrical terms, as illustrated in figure 5-8 for the case where M=2 and N=4:

1) Target: desired output vector:

$$d = \begin{pmatrix} d(M-1) & d(M-2) & \cdots & d(N-1) \end{pmatrix}^T$$

2) Candidates: computed output vector and its range:

$$y \equiv \begin{pmatrix} y(M-1) \\ y(M) \\ \vdots \\ y(N-1) \end{pmatrix} = \begin{pmatrix} x(M-1) & x(M-2) & \cdots & x(0) \\ x(M) & x(M-1) & \cdots & x(1) \\ \vdots & \vdots & \vdots & \vdots \\ x(N-1) & x(N-2) & \cdots & x(N-M) \end{pmatrix} \cdot \begin{pmatrix} w(0) \\ w(1) \\ \vdots \\ w(M-1) \end{pmatrix} \equiv X \cdot w$$

That is, the output vector is a linear combination of some input vectors. In other words, the output can be a vector anywhere in the subspace spanned by the M input vectors.

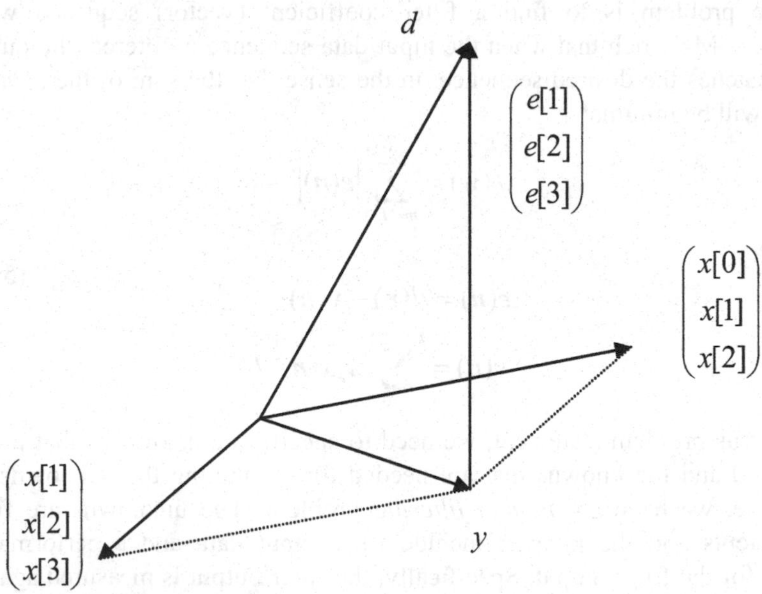

Figure 5-8. Geometric view of the least-squares optimal filtering problem

3) Metric: the "distance" between the desired vector and the computed vector: $L_{ss} = \|d - y\|^2$.

We can show that the desired weight vector w can be obtained by solving the following equation:

$$X^* X \cdot w = X^* \cdot d,$$

where * denotes Hermitian operation (i.e. combined transpose and conjugate operation).

When X has full column rank, the solution w is given by

$$\boxed{w = \left(X^* X\right)^{-1} X^* \cdot d} \qquad (5.18)$$

This is called the normal equation. This shows that for any model structure where the output is a linear function of the input, the optimal weight vector can be determined by this equation.

Now let us derive the normal equation another way.

The problem is to estimate unknown parameters

w_{0k}^*, $k = 1, 2, \ldots\ldots\ldots M$, given the data sequence

$\{x(n), d(n)\}$ $\quad n = 1, 2, \ldots\ldots\ldots N$ - length of data

d(n) is generated by $d(n) = \sum_{k=1}^{M} w_{0k}^* x(n-k+1) + e_0(n)$ where

w_{0k}^* are the optimal weights and $e_0(n) =$ measurement error which is statistical in nature with properties

$$E[e_0(n)] = 0 \forall n$$

$$E[e_0(n) e_0^*(k)] = \begin{cases} \sigma^2, & \text{for n} = k \\ 0, & \text{for n} \neq k \end{cases}$$

The solution is to use the linear transversal filter model to choose the filter weights w_k^* $(k = 1, 2, \ldots\ldots\ldots M)$ so as to minimize J(w) where

$$J(\mathbf{w}) = \sum_{n=i_1}^{i_2} |e(n)|^2 \qquad \mathbf{w} = \begin{pmatrix} w_1 \\ w_2 \\ \vdots \\ w_M \end{pmatrix}$$

where

$$e(n) = d(n) - y(n) \text{ is the estimation error}$$

$$y(n) = \sum_{k=1}^{M} w_k^* x(n-k+1) \text{ is the filter output}$$

We can choose limits i_1 and i_2 on J(**w**) in 4 different ways:
1. covariance method,
2. auto-correlation method,
3. pre-windowing method,
4. post-windowing method.

Using the covariance method $J(w_1, w_2, ...w_m) = \sum_{n=M}^{N} | e(n) |^2$

$$\mathbf{w} = \begin{pmatrix} w_1 \\ w_2 \\ \vdots \\ w_M \end{pmatrix} \qquad \mathbf{x}(n) = \begin{pmatrix} x(n) \\ x(n-1) \\ \vdots \\ x(n-M+1) \end{pmatrix}, \quad M \le n \le N$$

$$e(n) = d(n) - \mathbf{w}^H \mathbf{x}(n), \qquad M \le n \le N \quad \text{[N-M+1 equations]}$$

Let us define the estimation error vector as

$$\underline{\mathbf{e}}^H = [e(M), e(M+1), ..., e(N)]$$

Define the desired response vector

$$\underline{\mathbf{d}}^H = [d(M), d(M+1), ..., d(N)]$$

$$\mathbf{e}^H = \mathbf{d}^H - \mathbf{w}^H U^H \quad \text{where}$$

$$U^H = [\mathbf{x}(M), \mathbf{x}(M+1), ..., \mathbf{x}(N)]$$

$$\mathbf{e} = \mathbf{d} - U\mathbf{w}$$

U is an (N-M+1)xM data matrix (for the covariance method). The size of U depends on the method used. Next, we display the matrix U for all four methods.

Covariance method

$$i_1 = M \ , \ i_2 = N$$

This choice means we have made no assumptions about data outside interval $(1, N)$

$$U = \begin{pmatrix} x(M) & x(M+1)\cdots & x(N) \\ x(M-1) & x(M)\cdots & x(N-1) \\ \vdots & \vdots & \vdots \\ x(1) & x(2)\cdots\cdots & x(N-M+1) \end{pmatrix} \updownarrow M$$

$$\longleftarrow \quad N-M+1 \quad \longrightarrow$$

Auto-Correlation method

$$i_1 = 1 \ , \ i_2 = N+M-1$$

This choice means we assume data prior to time $i = 1$ and after $i = N$ are zero.

$$U = \begin{pmatrix} x(1) & x(2)\cdots & x(M) & x(M+1)\cdots & x(N) & 0\cdots\cdots & 0 \\ 0 & x(1)\cdots & x(M-1) & x(M)\cdots & x(N-1) & x(N)\cdots & 0 \\ \vdots & \vdots & \vdots & \vdots & \vdots & \vdots & \vdots \\ 0 & 0\cdots\cdots & x(1) & x(2)\cdots & x(N-M+1) & x(N-M)\cdots & x(N) \end{pmatrix} \updownarrow M$$

$$\longleftarrow \quad \text{N+M-1} \quad \longrightarrow$$

Pre-windowing method

$$i_1 = 1 , \ i_2 = N$$

This choice means we assume data prior to time $i = 1$ are zero.
There are no assumptions about data after time $i = N$.

$$U = \begin{pmatrix} x(1) & x(2)\cdots & x(M) & x(M+1)\cdots & x(N) \\ 0 & x(1)\cdots & x(M-1) & x(M)\cdots & x(N-1) \\ \vdots & \vdots & \vdots & \vdots & \vdots \\ 0 & 0\cdots\cdots & x(1) & x(2)\cdots & x(N-M+1) \end{pmatrix} \updownarrow \text{M}$$

$$\longleftarrow \quad \text{N} \quad \longrightarrow$$

Post-windowing method

$$i_1 = M , \ i_2 = N+M-1$$

This last choice means no assumptions are made about data prior to time $i=1$. We assume data after time $i=N$ are zero.

$$U=\begin{pmatrix} x(M) & x(M+1)\cdots & x(N) & 0\cdots & 0 \\ x(M-1) & x(M)\cdots & x(N-1) & x(N)\cdots & \vdots \\ \vdots & \vdots & \vdots & \vdots & 0 \\ x(1) & x(2)\cdots\cdots & x(N-M+1) & x(N-M)\cdots & x(N) \end{pmatrix} \Big\updownarrow M$$

$$\longleftarrow \quad N \quad \longrightarrow$$

Properties of the least-square/normal equation solutions

1. The least squares estimate w_0 is unique if the null space of the data matrix U is zero.

For the covariance method, U= LxM matrix where L= N-M+1.
The null space of U is the space of all vectors $\mathbf{x}$ such that U$\mathbf{x}$=0

$\Rightarrow$ at least $N-M+1\geq M$ columns are linearly independent

$\Rightarrow$ $U\mathbf{w}=\mathbf{d}$ is over determined (more equations than unknowns)

$\Rightarrow U^H U$ is non-singular

$\Rightarrow \mathbf{w_0}=(U^H U)^{-1}U^H\mathbf{d}$ is unique.

2. At the optimum point, $\mathbf{w_0}$ the minimum mean square error is given by

$$\begin{aligned} J_{\min} &= \mathbf{e}^H_{\min}\mathbf{e}_{\min} \\ &= [\mathbf{d}^H-\mathbf{w_0}^H U^H][\mathbf{d}-U\mathbf{w_0}] \\ &= \mathbf{d}^H\mathbf{d}-\mathbf{w_0}^H U^H\mathbf{d}-\mathbf{d}^H U\mathbf{w_0}+\mathbf{w_0}^H U^H U\mathbf{w_0} \end{aligned}$$

Recall $U^H U w_0 = U^H d$ (normal equation)

$$J_{\min}=\mathbf{d}^H\mathbf{d}-\mathbf{d}^H U\mathbf{w_0}$$

$$=\mathbf{d}^H\mathbf{d}-\mathbf{d}^H U[(U^H U)^{-1}U^H\mathbf{d}]$$

$$\sigma_d=\mathbf{d}^H\mathbf{d}=\sum_{n=M}^{N}|d(n)|^2$$

$$J_{\min}>0$$

$J_{\min} = \sigma_d$ when $e_0(n) = 0 \quad \forall \quad M \le n \le N$ which is an

Under-determined linear least squares estimation problem.

$$\Rightarrow \sum_{n=M}^{N} x(n-k+1)e_{\min}^{*}(n) = 0 \ , \qquad k = 1,2,.........M$$

$$e_{\min}(n)$$

only when $U^H \mathbf{e}_{\min} = \mathbf{0} \quad \Rightarrow \quad \mathbf{w} = \mathbf{w_0}$

$\mathbf{d_0} = U\mathbf{w_0}$ = least square estimate of **b** (desired response vector)

$$U^H \mathbf{e}_{\min} = \mathbf{0}$$

$$w^H U^H \mathbf{e}_{\min} = d^H \mathbf{e}_{\min}^H = \mathbf{0} \quad \Rightarrow \quad \mathbf{w} = \mathbf{w}^{\wedge}$$

$$J(\mathbf{w}) = \mathbf{e^H e}$$

$$= (\mathbf{d}^H - \mathbf{w}^H U^H)(\mathbf{d} - U\mathbf{w})$$

$$= \mathbf{d}^H\mathbf{d} - \mathbf{d}^H U\mathbf{w} - \mathbf{w}^H U^H \mathbf{d} + \mathbf{w}^H U^H U\mathbf{w}$$

$$\frac{\partial J(\mathbf{w})}{\partial \mathbf{w}} = -2U^H d + 2U^H U\mathbf{w}$$

$$= -2U^H \mathbf{e}$$

let $\mathbf{w}_0$ be the value of $\mathbf{w}$ for which

$$J(\mathbf{w}) = \text{minimum} \qquad \text{or} \qquad \frac{\partial J}{\partial \mathbf{w}} = 0$$

$\Rightarrow U^H U\mathbf{w_0} = U^H d$ - deterministic normal equation for LLSE problem

$$\boxed{\mathbf{w_0} = (U^H U)^{-1} U^H d} \qquad (5.19)$$

3. Orthogonality Conditions

$$\mathbf{e}_{\min} = \mathbf{d} - U\mathbf{w_0}$$

$$U^H \mathbf{e}_{\min} = \mathbf{0} \quad \text{(orthogonality of A \& } \mathbf{e}_{\min}\text{)}$$

Properties of Least-Squares Estimates

1. $\hat{w}$ is unbiased provided $\{e_0(i)\}$, the measurement error process, has zero mean.
2. When $\{e_0(i)\}$ is white, zero mean and variance σ^2, the covariance matrix of the least-squares estimate $\hat{w}$ equals $\sigma^2\phi^{-1}$.
3. When $\{e_0(i)\}$ is white, zero mean, the least square estimate $\hat{w}$ is the best, linear, unbiased estimate (BLUE).
4. When $\{e_0(i)\}$ is white, Gaussian, zero-mean, the least squares estimate $\hat{w}$ achieves the Cramer-Rao lower bound for unbiased estimates.

Meanings of $U^H\mathbf{d}$ and $U^H U$

1. $U^H\mathbf{d}$ = deterministic cross-correlation vector

$$= \theta = \sum_{n=M}^{N} \mathbf{u}(n)d^*(n)$$

$$\mathbf{u}(n)^T = [x(n)\ \ x(n-1)\ldots\ldots\ldots\ldots x(n-M+1)] \qquad M \le n \le N$$

$$\theta = \begin{pmatrix} \theta(0) \\ \theta(1) \\ \vdots \\ \theta(M-1) \end{pmatrix}$$

$$\theta(t) = \sum_{n=M}^{N} x(n-t)d^*(n) \qquad 0 \le t \le M-1$$

2. $U^H U$ = deterministic correlation matrix

$$\phi = \sum_{n=M}^{N} \mathbf{u}(n)\mathbf{u}^H(n)$$

$$\phi(t,k) = \sum_{i=M}^{N} x(i-t)x^{*}(i-k) \qquad 0 \le t\,,\ k \le M-1$$

$$\boxed{\phi w^{\wedge} = \theta} \tag{5.20}$$

Deterministic Normal Equations

$$J_{\min} = \sigma_d - \theta^H w_0$$

$$= \sigma_d - \theta^H \phi^{-1} \theta$$

Properties of ϕ

1. ϕ is Hermitian.

$$\Rightarrow \quad \phi^H = \phi$$

Proof:

$$\phi = U^H U \ , \ \phi^H = U^H U$$

2. ϕ is non-negative definite.

$$\mathbf{y}^H \phi \mathbf{y} \ge 0 \quad \forall \quad \mathbf{y} = \text{Mx1 vector}$$

Proof:

$$\mathbf{y}^H \phi \mathbf{y} = \sum_{n=M}^{N} \mathbf{y}^H \mathbf{u}(n) \mathbf{u}^H(n) \mathbf{y}$$

$$= \sum_{n=M}^{N} [\mathbf{y}^H \mathbf{u}(n)][\mathbf{y}^H \mathbf{u}(n)]^H$$

$$= \sum_{n=M}^{N} \left| \mathbf{y}^H \mathbf{u}(n) \right|^2 \ge 0$$

$$\mathbf{y}^H \phi \mathbf{y} > 0 \quad \Rightarrow \quad \phi^{-1} \ \text{exists}$$

3. Eigenvalues of ϕ are all real and non-negative.

Proof: Follows from properties 1 & 2.

4. ϕ is a product of 2 rectangular Toeplitz matrices.

Proof: ϕ is clearly non-Toeplitz.

$$\phi = U^H U$$

$$U^H = \begin{pmatrix} x(M) & x(M+1)\cdots & x(N) \\ x(M-1) & x(M)\cdots & x(N-1) \\ \vdots & \vdots & \\ x(1) & x(2)\cdots & x(N-M+1) \end{pmatrix} \updownarrow M$$

$$\longleftarrow \text{N-M+1} \longrightarrow$$

U^H = rectangular Toeplitz

U = rectangular Toeplitz

Relationship of the Normal Equations to Nonlinear System Modeling

The normal equations can be used to fit a polynomial to a sequence of measurements. For example, given N measurements of the input signal, $x_1, x_2, \ldots, x_N$ and an output z, we can fit a q th order polynomial to relate the input x to the output z. Define y_j as:

$$y_j = c^{(0)} + c^{(1)} x_j + c^{(2)} x_j^{\ 2} + c^{(3)} x_j^{\ 3}, j = 1, \ldots, N$$

$$z = y + \nu$$

where ν is a zero-mean additive white Gaussian noise sequence.

Define the matrix X as the matrix with N (# of data inputs) rows and Q=q+1 (# of parameters) columns.

$$X = \begin{pmatrix} 1 & x_1 & x_1^2 & \cdots & x_1^q \\ 1 & x_2 & x_2^2 & \cdots & x_2^q \\ 1 & x_3 & x_3^2 & \cdots & x_3^q \\ \vdots & \vdots & \vdots & \ddots & \vdots \\ 1 & x_N & x_N^2 & \cdots & x_N^q \end{pmatrix}$$

Then the solution for z is

$$z = Xw + v$$

where $w = [c^{(0)}, c^{(1)}, c^{(2)}, c^{(3)}, ..., c^{(q)}]$ is the parameter vector and v is the additive noise.

5.5 Adaptive Filters (RLS-Based Algorithms)

The solution to the method of least-squares is now made recursive, resulting in the recursive least squares (RLS) algorithm.

We can use a multiple linear regression model to generate a desired signal

$$d(i) = \sum_{k=0}^{M-1} w_{0k}^* x(n-k) + e_0(n) \tag{5.21}$$

where w_{0k} are the unknown parameters of the model and $e_0(n)$ represents the measurement error to which the statistical nature of the phenomenon is attributed. It is assumed to be white noise of zero mean and variance σ^2.

A linear transversal (FIR) filter excited by white noise is used to model the generation of the desired response.

The goal is to minimize the cost function

$$J(n) = \sum_{i=1}^{n} \beta(n,i) \mid e(i) \mid^2 \tag{5.22}$$

$$0 < \beta(n,i) \leq 1 \qquad i = 1, 2, ...n$$

$$e(i) = d(i) - y(i)$$

$$= d(i) - \mathbf{w}^H(n)\mathbf{u}(i)$$

$$u(i)^T = [u(i)u(i-1)...........u(i-M+1)]$$

$$= \text{input vector at time i}$$

$$w(n)^T = [w_1(n)w_2(n)...............w_M(n)]$$

$$= \text{weight-vector at time n}$$

Typically we use $\beta(n,i) = \lambda^{n-i}$, $i = 1,2,............n$, resulting in

$$J(w(n)) = \sum_{i=1}^{n} \lambda^{n-i} |e(i)|^2 \tag{5.23}$$

for the exponentially-weighted least squares algorithm. Note that n is bounded by $0 < n \leq N$, the maximum number of data used for the least squares.

$$0 < \lambda \leq 1 \text{ (for example, } \lambda = 0.9\text{)}$$

Choice of $\lambda = 1$ results in the least squares method of the last section.

$$\frac{1}{1-\lambda} = \text{ memory of the algorithm}$$

Here we have chosen the limits of the summation in one of the four different ways mentioned in the last section.

Recall that these are

1. covariance method: i1=M, i2=N,
2. auto-correlation method: i1=1, i2=N+M-1,
3. pre-windowing method: i1=1, i2=N,
4. post-windowing method: i1=M, i2=N+M-1.

Each of these selections indicates an implicit assumption about how to deal with the data prior to i1=1, or prior to i1=M, or after i2=N, or after i2=N+M-1.

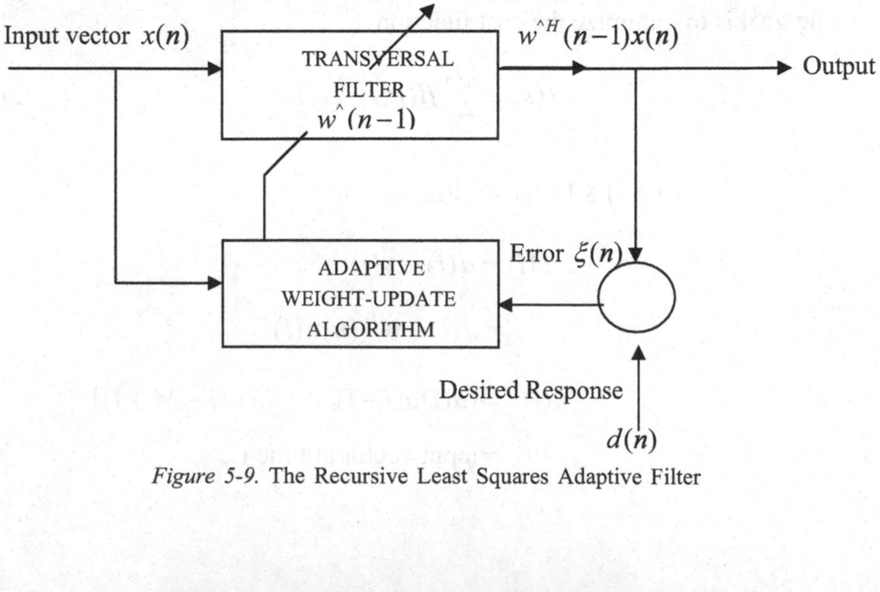

Figure 5-9. The Recursive Least Squares Adaptive Filter

We assume a transversal-filter (FIR filter) structure as shown in Figure 5.9. We are given the least-square estimate of the tap-weight vector of the filter at time (n-1). We need the estimate at time (n) based on arrival of new data at time (n). The RLS algorithm uses all the information contained in input data from time i=1 to time i=n (current time)—for the post-windowing method, for example—to determine the least-square solution in a *recursive* way.

Assume the exponential weighted least-squares (with post-windowing method), then the optimum weight vector satisfies

$$\phi(n)\mathbf{w}(n) = \theta(n) \quad \text{normal equations,} \tag{5.24}$$

where correlation matrix $\phi(n) = \sum_{i=1}^{n} \lambda^{n-i}\mathbf{u}(i)\mathbf{u}^H(i)$ MxM matrix

and cross-correlation vector $\theta(n) = \sum_{i=1}^{n} \lambda^{n-i}\mathbf{u}(i)d^*(i)$ Mx1 vector

$\phi(n), \theta(n)$ differ from that defined for least-squares by

1. λ^{n-i} weighting
2. use of prewindowing

$$\Rightarrow \quad i_1 = 1, \quad i_2 = n$$

$$\phi(n-1) = \sum_{i=1}^{n-1} \lambda^{n-1-i}\mathbf{u}(i)\mathbf{u}^H(i).$$

A recursive formulation for $\phi(n), \theta(n)$ is shown below:

$$\phi(n) = \lambda\phi(n-1) + \mathbf{u}(n)\mathbf{u}^H(n) \tag{5.25}$$

Similarly,

$$\theta(n) = \lambda\theta(n-1) + \mathbf{u}(n)d^*(n) \tag{5.26}$$

Matrix Inversion Lemma

If $A = B^{-1} + CD^{-1}C^H$,

then $A^{-1} = B - BC(D + C^H BC)^{-1}C^H B$.

Here, we substitute the following as the matrices A, B, C and D:

$$A = \phi(n)$$

$$B^{-1} = \lambda\phi(n-1)$$

$$C = u(n)$$

$$D = 1$$

We need to use the matrix inversion lemma to proceed.
A, B are two positive-definite, MxM matrices.
$D =$NxN , positive definite matrix.
$C =$ MxN matrix.
Here $M = M, N = 1$.
(See appendix 5A for the derivation).

If we substitute this in equation (5.25), using the assignments above gives:

$$\phi^{-1}(n) = \lambda^{-1}\phi^{-1}(n-1) - \frac{\lambda^{-2}\phi^{-1}(n-1)\mathbf{u}(n)\mathbf{u}^H(n)\phi^{-1}(n-1)}{1+\lambda^{-1}\mathbf{u}^H(n)\phi^{-1}(n-1)\mathbf{u}(n)} \qquad (5.27)$$

Let $P(n) = \phi^{-1}(n)$ (MxM matrix).

$$K(n) = \frac{\lambda^{-1}P(n-1)\mathbf{u}(n)}{1+\lambda^{-1}\mathbf{u}^H(n)P(n-1)\mathbf{u}(n)} \qquad \text{(Mx1 vector)}$$

Equation (5.27) becomes

$$P(n) = \lambda^{-1}P(n-1) - \lambda^{-1}K(n)\mathbf{u}^H(n)P(n-1) \qquad (5.28)$$

$$K(n) = \lambda^{-1}P(n-1)\mathbf{u}(n) - \lambda^{-1}K(n)\mathbf{u}^H(n)P(n-1)\mathbf{u}(n)$$

$$= [\lambda^{-1}P(n-1) - \lambda^{-1}K(n)\mathbf{u}^H(n)P(n-1)]\mathbf{u}(n)$$

$$= P(n)\mathbf{u}(n)$$

$$\Rightarrow \quad \boxed{K(n)=\phi^{-1}(n-1)\mathbf{u}(n)} \quad \text{gain vector} \tag{5.29}$$

Now recall weight vector

$$w(n)=\phi^{-1}(n)\theta(n)$$

$$w(n)=P(n)\theta(n)$$

$$=\lambda P(n)\theta(n-1)+P(n)\mathbf{u}(n)d^{*}(n)$$

$$w(n)=\lambda[\lambda^{-1}P(n-1)-\lambda^{-1}k(n)\mathbf{u}^{H}P(n-1)]\theta(n-1)+P(n)\mathbf{u}(n)d^{*}(n)$$

$$=P(n-1)\theta(n-1)-K(n)u^{H}(n)P(n-1)\theta(n-1)+P(n)\mathbf{u}(n)d^{*}(n)$$

$$w(n)=w(n-1)-K(n)\mathbf{u}^{H}(n)w(n-1)+P(n)\mathbf{u}(n)d^{*}(n)$$

$$=w(n-1)+K(n)[d^{*}(n)-\mathbf{u}^{H}(n)w(n-1)]$$

$\alpha(n)$ = innovation or prior estimation error

$\alpha(n)=d(n)-w^{H}(n-1)\mathbf{u}(n)$

Estimation of $d(n)$ based on $w(n-1)$ old, least-squares estimate

$e(n)=d(n)-w^{H}(n)\mathbf{u}(n)$ = a posteriori estimation error

Estimate of $d(n)$ based on $w(n)$, current least-square estimate

$$\boxed{w(n)=w(n-1)+K(n)\alpha^{*}(n)} \tag{5.30}$$

This is the RLS algorithm weight update equation.

Initial Conditions

1. Choose $P(0)$ so that $\phi^{-1}(0)$ exists.

$$\Rightarrow \text{use data for } -n_0\le i\le 0$$

$$P(0)=[\sum_{i=-n_0}^{0}\lambda^{-i}\mathbf{u}(i)\mathbf{u}^{H}(i)]^{-1}$$

Redefine

$$\phi(n) = \sum_{i=1}^{n} \lambda^{n-i} \mathbf{u}(i)\mathbf{u}^H(i) + \delta\lambda^n I$$

where I = MxM identity

δ = small positive constant

Note

This redefinition of $\phi(n)$ does not affect the RLS algorithm derived earlier

$$\phi(0) = \delta I \quad \Rightarrow \quad P(0) = \delta^{-1} I$$

2. Choose $w(0) = \mathbf{0}$ (null Mx1 vector)

Summary of the RLS Algorithm

Initialize the algorithm by setting

$$w^{\wedge}(0) = 0$$

$$P(0) = \delta^{-1} I$$

and

δ = small positive constant for high SNR
large positive constant for low SNR

For each instant of time, n=1,2… compute

$$\pi(n) = P(n-1)u(n)$$

$$k(n) = \frac{\pi(n)}{\lambda + u^H(n)\pi(n)}$$

$$\xi(n) = d(n) - w^{\wedge H}(n-1)u(n)$$

$$w^{\wedge}(n) = w^{\wedge}(n-1) + k(n)\xi^*(n)$$

and

$$P(n) = \lambda^{-1}P(n-1) - \lambda^{-1}k(n)u^H(n)P(n-1)$$

5.6 Summary

In summary, we have presented a brief introduction to the vast field of adaptive signal processing. We started with the ideal Wiener filter (not to be confused with the nonlinear Wiener model) for linear estimation. Then we discussed the LMS-based algorithms. Later we introduced linear least squares and the resulting RLS-based algorithms. These are the most common algorithms amongst many used in the field.

5.7 Appendix 5A

ABCD (INVERSION) LEMMA: INVERSION OF [A+BCD]

$$(A+BCD)^{-1} = A^{-1} - A^{-1}B\left(C^{-1} + DA^{-1}B\right)^{-1}DA^{-1}$$

ABCD lemma is easy to prove:

$$\begin{aligned}
&(A+BCD)\left(A^{-1} - A^{-1}B\left(C^{-1} + DA^{-1}B\right)^{-1}DA^{-1}\right) \\
&= I - B\left(C^{-1} + DA^{-1}B\right)^{-1}DA^{-1} + BCDA^{-1} - BCDA^{-1}B\left(C^{-1} + DA^{-1}B\right)^{-1}DA^{-1} \\
&= I + B\left[-I_1 + C\left(C^{-1} + DA^{-1}B\right) - CDA^{-1}B\right]\left(C^{-1} + DA^{-1}B\right)^{-1}DA^{-1} \\
&= I
\end{aligned}$$

where I and I_1 are respectively of the same dimension as A and C.

Although the proof is easy, one might be tempted to ask how this formula can be invented in the first place. The following is one way: Our strategy is first to solve a related simpler problem and then use the result to solve the original problem.

First we will assume that A and C are identity matrices of their respective dimensions. Just as we can prove $(1-r)^{-1} = \sum_{n=0}^{\infty} r^n$ for r < 1, we can show $(I-R)^{-1} = \sum_{n=0}^{\infty} R^n$ when all eigenvalues of R are smaller than 1. Since we can verify the formula after getting it, at this stage we will assume that the relevant matrices have all the properties we need to proceed.

Thus we have:

$$\begin{aligned}(I-BD)^{-1} &= \sum_{n=0}^{\infty}(BD)^n \\ &= I+\sum_{n=1}^{\infty}(BD)^n \\ &= I+B\sum_{n=1}^{\infty}(DB)^{n-1}D \\ &= I+B(I_1-DB)^{-1}D.\end{aligned}$$

Proceeding to tackle the general case, we have

$$(A+BCD)^{-1}=[A(I+B_1D)]^{-1}$$

where we used the notation

$$B_1 \equiv A^{-1}BC\,.$$

Using the formula we obtained for the simpler case, we get

$$\begin{aligned}(A+BCD)^{-1} &= \left[I-B_1(I_1+DB_1)^{-1}D\right]A^{-1} \\ &= A^{-1}-A^{-1}BC\left(I_1+DA^{-1}BC\right)^{-1}DA^{-1} \\ &= A^{-1}-A^{-1}BC\left[\left(C^{-1}+DA^{-1}B\right)C\right]^{-1}DA^{-1} \\ &= A^{-1}-A^{-1}B\left(C^{-1}+DA^{-1}B\right)^{-1}DA^{-1}\end{aligned}$$

Chapter 6

NONLINEAR ADAPTIVE SYSTEM IDENTIFICATION BASED ON VOLTERRA MODELS

Algorithms based on the Volterra and bilinear models

Introduction

The Volterra series model has become quite popular in adaptive nonlinear filtering circles in the last few years. A linear system can be completely described by the unit impulse response. The Volterra series representation is an extension of linear system theory. This extension shows the highly complex nature of nonlinear filtering. Consequently, many researchers have restricted themselves to certain low-order systems which are mostly based on the Volterra model.

In this chapter, given the background in chapters 3 and 5, we develop a nonlinear LMS adaptive algorithm which is based on the discrete Volterra model. This nonlinear LMS adaptive method can be seen as an extension of the linear LMS algorithm. The merit of this approach is that it keeps most of the linear LMS properties but still has a reasonably good convergence rate. The performance analysis is also more tractable, which is seldom true for most nonlinear adaptive algorithms. Furthermore, the extension to the general higher order is straightforward and it can even apply to the least mean fourth (LMF) family of adaptive algorithms.

One of the first applications of nonlinear adaptive system identification was for a second-order polynomial nonlinear application in the papers by Koh and Powers (Koh 1983, Koh 1985). These papers focused on using a Volterra series polynomial model to model the nonlinear system and opened up the possibility of realizing adaptive filters for a nonlinear system in this fashion.

Also in this chapter, we introduce the bilinear model for nonlinear systems and apply the LMS-type adaptive algorithm to its coefficients for adaptation. Later, we use the RLS-type algorithm for adaptation of the coefficients of the truncated Volterra series model. Computer simulation results are discussed.

6.1 LMS Algorithm for Truncated Volterra Series Model

The development of a gradient-type LMS adaptive algorithm for truncated Volterra series nonlinear models follows a similar method of development as for linear systems. This was covered in chapter 5.

The truncated *P*th order Volterra series expansion is

$$y[n] = h_0 + \sum_{m_1=0}^{N-1} h_1[m_1]x[n-m_1] + \sum_{m_1=0}^{N-1}\sum_{m_2=0}^{N-1} h_2[m_1,m_2]x[n-m_1]x[n-m_2] + \cdots$$
$$+ \sum_{m_1=0}^{N-1}\sum_{m_2=0}^{N-1}\cdots\sum_{m_p=0}^{N-1} h_p[m_1,m_2,\ldots,m_p]x[n-m_1]x[n-m_2]\cdots x[n-m_p], \quad (6.1)$$

Assuming h0=0 and p=2, the weight vector is

$$H[n] = \{h_1[0;n], h_1[1;n], \ldots, h_1[N-1;n], h_2[0,0;n], h_2[0,1;n], \ldots,$$
$$h_2[0,N-1;n], h_2[1,1;n], \ldots, h_2[N-1,N-1;n]\}^T \quad (6.2)$$

The input vector is

$$X[n] = \{x[n], x[n-1], \ldots, x[n-N+1], x^2[n], x[n]x[n-1], \ldots,$$
$$x[n]x[n-N+1], x^2[n-1], \ldots, x^2[n-N+1]\}^T \quad (6.3)$$

Linear and quadratic coefficients are updated separately by minimizing the instantaneous square of the error

$$J(n) = e^2(n) \quad \text{where} \quad e(n) = d(n) - \hat{d}(n) \quad (6.4)$$

where $y(n) = \hat{d}(n)$ is the estimate of $d(n)$. This results in the update equations

$$h_1[m_1;n+1] = h_1[m_1;n] - \frac{\mu_1}{2}\frac{\partial e^2(n)}{\partial h_1[m_1;n]} = h_1[m_1;n] + \mu_1 e(n)x(n-m_1) \quad (6.5)$$

$$h_2[m_1,m_2;n+1] = h_2[m_1,m_2;n] - \frac{\mu_1}{2}\frac{\partial e^2(n)}{\partial h_2[m_1,m_2;n]}$$
$$= h_2[m_1,m_2;n] + \mu_2 e(n)x(n-m_1)x(n-m_2) \quad (6.6)$$

where μ_1 and μ_2 are step-sizes used to control speed of convergence and ensure stability of the filter.

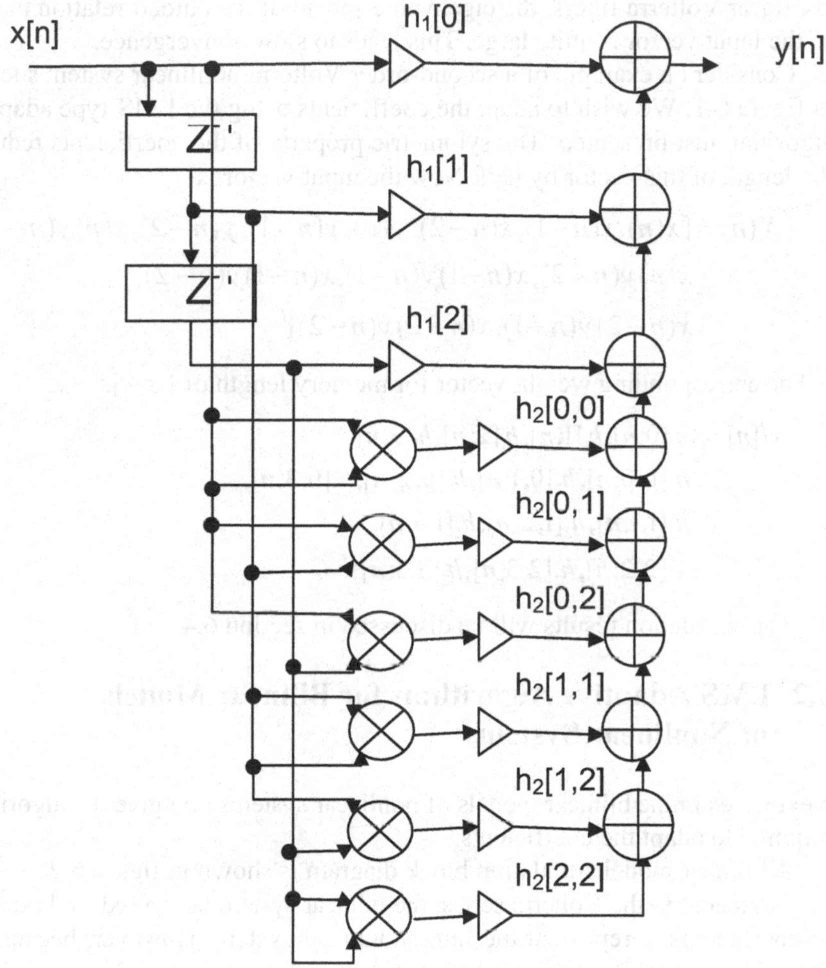

Figure 6-1. Second order Volterra series model used for adaptation N=3, p=2

Using the weight vector notation, $H[n]$, we can combine the two update equations into one as the coefficient update equation

$$e[n] = d[n] - H^T[n]X[n]$$
$$H[n+1] = H[n] + \mu X[n]e[n]$$

where μ is chosen such that $0 < \mu_1,\ \mu_2 < \dfrac{2}{\lambda_{\max}}$ and $\lambda_{\max}$ is the maximum eigenvalue of the autocorrelation matrix of the input vector $X[n]$. For nonlinear Volterra filters, the eigenvalue spread of the autocorrelation matrix of the input vector is quite large. This leads to slow convergence.

Consider an example of a second-order Volterra nonlinear system such as in figure 6-1. We wish to adapt the coefficients using the LMS-type adaptive algorithm just presented. The symmetric property of the coefficients reduces the length of this vector by half. Now the input vector is

$$\begin{aligned} X(n) = [&x(n), x(n-1), x(n-2), y(n), y(n-1), y(n-2), x(n)y(n-1), \\ &x(n)y(n-2), x(n-1)y(n-1), x(n-1)y(n-2), \\ &x(n-2)y(n-1), x(n-2)y(n-2)]^T \end{aligned}$$

The corresponding weight vector for memory length of N=4 is

$$\begin{aligned} H[n] = \{&h_1[0;n], h_1[1;n], h_1[2;n], h_1[3;n], \\ &h_2[0,0;n], h_2[0,1;n], h_2[0,2;n], h_2[0,3;n], \\ &h_2[1,1;n], h_2[1,2;n], h_2[1,3;n], \\ &h_2[2,2;n], h_2[2,3;n], h_2[3,3;n]\}^T \end{aligned}$$

The simulation results will be discussed in section 6.4.

6.2 LMS Adaptive Algorithms for Bilinear Models of Nonlinear Systems

Next we examine bilinear models of nonlinear systems and give the algorithm required to adapt the coefficients.

A bilinear model simulation block diagram is shown in figure 6-2.

Compared to the Volterra series, the bilinear system uses a reduced number of coefficients to represent the same nonlinear system. However, because of the feedback, stability may be an issue.

In general, we assume the output signal is of the form

$$y[n] = \sum_{i=1}^{M} P_i(y[n-1], y[n-2], \ldots, y[n-N+1], x[n], x[n-1], \ldots, x[n-N+1])$$

where $P_i(.)$ is an i'th order polynomial of the signals within the brackets. The bilinear system is a simple form of this general form, where the first- and second-order polynomials are used.

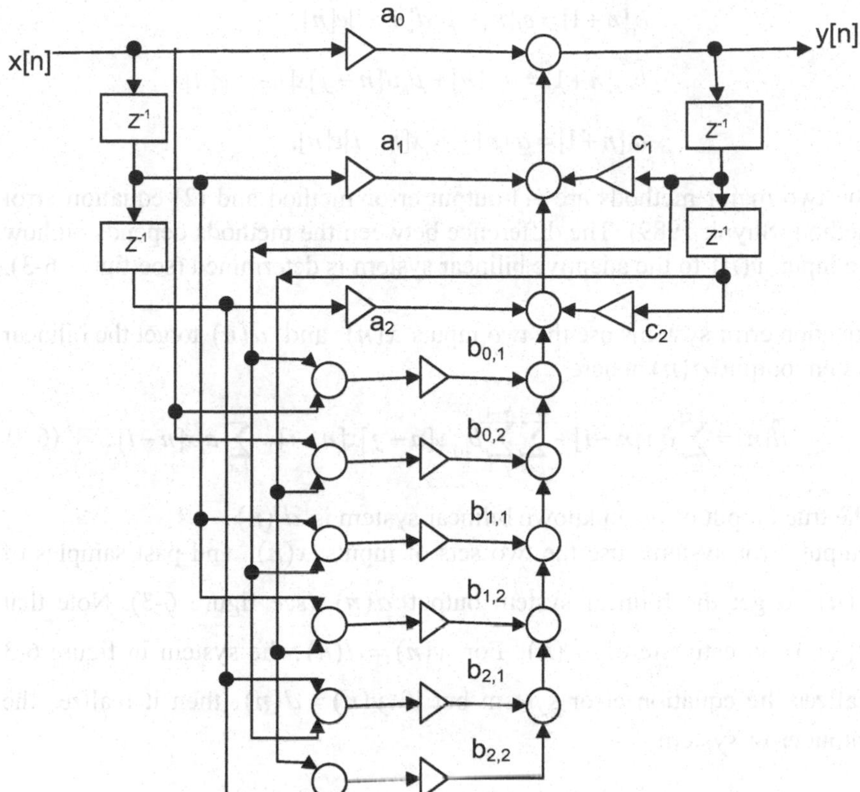

Figure 6-2. Implementation block diagram of a simple bilinear system

For example, when N=3, we have the following bilinear system:

$$y[n] = \sum_{i=1}^{N-1} c_i y[n-i] + \sum_{i=0}^{N-1}\sum_{j=1}^{N-1} b_{i,j} y[n-j]x[n-i] + \sum_{i=0}^{N-1} a_i x[n-i] \qquad (6.7)$$

A simple bilinear system can be adapted by either of the two methods shown in figure 6-3.

The weight update coefficient vector here is

$$W(n) = [a_0(n), a_1(n), a_2(n), c_0(n), c_1(n), c_2(n), \\ b_{0,1}(n), b_{0,2}(n), b_{1,1}(n), b_{1,2}(n), b_{2,1}(n), b_{2,2}(n)]^T$$

The weight estimate update equations are (with hats on the variables):

$$\hat{c}_i[n+1] = \hat{c}_i[n] + \mu_c d[n-i]e[n]$$

$$\hat{b}_{i,j}[n+1] = \hat{b}_{i,j}[n] + \mu_b d[n-j]x[n-i]e[n]$$

$$\hat{a}_i[n+1] = \hat{a}_i[n] + \mu_a x[n-i]e[n].$$

The two major methods are (1) output error method and (2) equation error method (Shynk 1989). The difference between the methods depends on how the input $y(n)$ to the adaptive bilinear system is determined (see figure 6-3).

Equation error systems use the two inputs $x(n)$ and $d(n)$ to get the bilinear system output $\tilde{d}(n)$ where

$$\hat{d}[n] = \sum_{i=1}^{N-1} \hat{c}_i y[n-i] + \sum_{i=0}^{N-1}\sum_{j=1}^{N-1} \hat{b}_{i,j} y[n-j]x[n-i] + \sum_{i=0}^{N-1} \hat{a}_i x[n-i]. \quad (6.8)$$

The true output of the unknown bilinear system is $d'(n)$.
Output error systems use the two sets of inputs $x(n)$ and past samples of $\hat{d}(n)$ to get the bilinear system output $\hat{d}(n)$ (see figure 6-3). Note that $\hat{d}(n)$ is an estimate of $d'(n)$. For $y(n) = d(n)$, the system in figure 6-3 realizes the equation error system but if $y(n) = \hat{d}(n)$, then it realizes the output error system.

With the presence of additive noise in $d(n)$, the resulting equation error estimates will be biased. However, the output error estimates are unbiased (or close to unbiased) after convergence.

The product terms in the bilinear system lead to multiplicative noise here instead of the usual additive noise in the linear system. This complicates the convergence analysis of these algorithms.

Stability and convergence are critical issues in both equation error and output error methods. However, we see that even though the equation error methods have a unique minimum, the gradient-type algorithms may not converge to that correct minimum point due to the additive noise at the desired response signal. However, output error methods lead to performance surfaces which have many local minimum, and gradient-type algorithms can easily get stuck in a local optimum point without proper initialization.

Later in section 6.4, we use the adaptive weight update algorithms to simulate the performance of the bilinear adaptive filter.

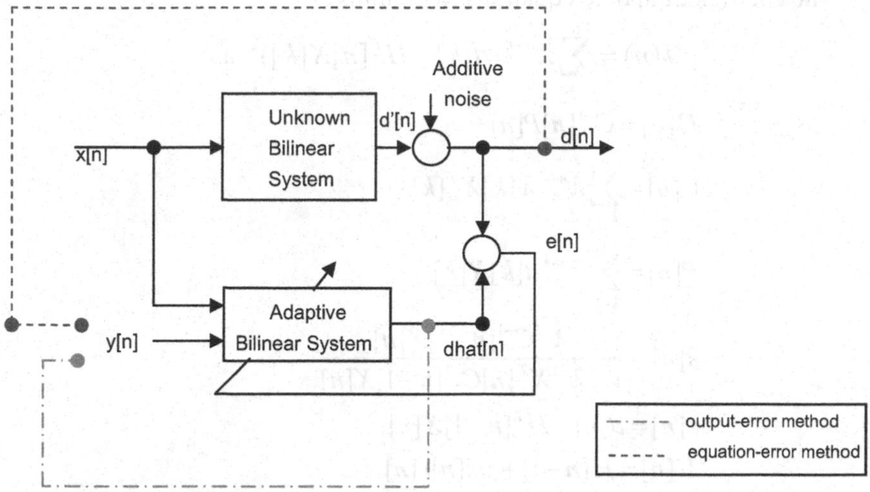

Figure 6-3. Equation error and output error methods for adaptation of the bilinear system

6.3 RLS Algorithm for Truncated Volterra Series Model

Here the RLS algorithm is applied to the truncated *P*th order Volterra series expansion given as:

$$y[n] = h_0 + \sum_{m_1=0}^{N-1} h_1[m_1]x[n-m_1] + \sum_{m_1=0}^{N-1}\sum_{m_2=0}^{N-1} h_2[m_1,m_2]x[n-m_1]x[n-m_2] + \cdots$$
$$+ \sum_{m_1=0}^{N-1}\sum_{m_2=0}^{N-1}\cdots\sum_{m_p=0}^{N-1} h_p[m_1,m_2,\ldots,m_p]x[n-m_1]x[n-m_2]\cdots x[n-m_p], \tag{6.9}$$

The weight vector is the same as before:

$$H[n] = \{h_1[0;n], h_1[1;n], \ldots, h_1[N-1;n], h_2[0,0;n],$$
$$h_2[0,1;n], \ldots, h_2[0,N-1;n], h_2[1,1;n], \ldots, h_2[N-1,N-1;n]\}^T \tag{6.10}$$

The input vector is the same as before:

$$X[n] = \{x[n], x[n-1], \ldots, x[n-N+1], x^2[n], x[n]x[n-1], \ldots,$$
$$x[n]x[n-N+1], x^2[n-1], \ldots, x^2[n-N+1]\}^T \tag{6.11}$$

The coefficient update equations to minimize

$$J(n) = \sum_{k=0}^{n} \lambda^{n-k} (d[k] - H^T[n]X[k])^2 \text{ are:}$$

$$H[n] = C^{-1}[n]P[n]$$

$$C[n] = \sum_{k=0}^{n} \lambda^{n-k} X[k]X^T[k]$$

$$P[n] = \sum_{k=0}^{n} \lambda^{n-k} d[k]X[k]$$

$$k[n] = \frac{\lambda^{-1}C^{-1}[n-1]X[n]}{1 + \lambda^{-1}X^T[n]C^{-1}[n-1]X[n]}$$

$$\varepsilon[n] = d[n] - H^T[n-1]X[n]$$

$$H[n] = H[n-1] + \mu k[n]\varepsilon[n]$$

$$C^{-1}[n] = \lambda^{-1}C^{-1}[n-1] - \lambda^{-1}k[n]X^T[n]C^{-1}[n-1]$$

$$e[n] = d[n] - H^T[n]X[n]$$

6.4 RLS Algorithm for Bilinear Model

Here the RLS algorithm is applied to the same bilinear model discussed in section 6.2 (Billings 1984, Ljung 1999). For example, when N=3, we have the following bilinear system:

$$y[n] = \sum_{i=1}^{N-1} c_i y[n-i] + \sum_{i=0}^{N-1}\sum_{j=1}^{N-1} b_{i,j} y[n-j]x[n-i] + \sum_{i=0}^{N-1} a_i x[n-i] \qquad (6.12)$$

This bilinear system can be adapted by either of the two methods shown in figure 6-3 but now using the extended (suboptimal) least-squares method.

The weight vector is:

$$\begin{aligned} H[n] = \{&\hat{c}_1[n], \hat{c}_2[n], \ldots, \hat{c}_{N-1}[n], \\ &\hat{b}_{0,1}[n], \ldots, \hat{b}_{N-1,N-1}[n], \\ &\hat{a}_0[n], \ldots, \hat{a}_{N-1}[n]\}^T \end{aligned}$$

The input vector is:

$$\begin{aligned} X[n] = \{&\hat{d}_{n-1}[n-1], \hat{d}_{n-2}[n-2], \ldots, \hat{d}_{n-N+1}[n-N+1], \\ &x[n]\hat{d}_{n-1}[n-1], \ldots, x[n-N+1]\hat{d}_{n-N+1}[n-N+1], \\ &x[n], x[n-1], \ldots, x[n-N+1]\}^T \end{aligned}$$

Note that here, $\hat{d}_k[l]$ represents the estimate of the desired response at time l given the adaptive filter coefficients at time k. At any time n, we recursively minimize

$$J(n) = \sum_{k=0}^{n} \lambda^{n-k} (d[k] - H^T[n]X[k])^2 \tag{6.13}$$

to get the adaptive filter coefficients at time n.

The coefficient update equations are:

$$\begin{aligned}
&\hat{d}_n[n] = H^T[n]X[n] \\
&H[n] = C^{-1}[n]P[n] \\
&C[n] = \sum_{k=0}^{n} \lambda^{n-k} X[k]X^T[k] \\
&P[n] = \sum_{k=0}^{n} \lambda^{n-k} d[k]X[k] \\
&k[n] = \frac{\lambda^{-1}C^{-1}[n-1]X[n]}{1+\lambda^{-1}X^T[n]C^{-1}[n-1]X[n]} \\
&\varepsilon[n] = d[n] - H^T[n-1]X[n] \\
&H[n] = H[n-1] + \mu k[n]\varepsilon[n] \\
&C^{-1}[n] = \lambda^{-1}C^{-1}[n-1] - \lambda^{-1}k[n]X^T[n]C^{-1}[n-1] \\
&e[n] = d[n] - H^T[n]X[n]
\end{aligned}$$

We note that this is the suboptimal (extended) least-squares solution and it is not really an exact least-squares solution. This is because the exact least-squares solution requires that we minimize the cost function $J(n)$ using the estimation error values $e_n[k] = d[k] - H^T[k]X[k]$ instead of $e[k] = d[k] - H^T[n]X[k]$, which uses $H[n]$ at current time n. Therefore the solution at time n depends on prior solutions implicitly, hence it is suboptimal least-squares.

6.5 Computer Simulation Examples

Example 1:

The first example is a simulation of figure 6-1 using both the LMS-type and RLS-type adaptive filters. The chosen coefficient values are:

$$H[n] = \{-0.78 - 1.48, -1.39, 0.04, \\ 0.54, 3.72, 1.86, -0.76, \\ -1.62, 0.76, -0.12, \\ 1.41, -1.52, -0.13\}^T$$

For the LMS-type algorithm, we use the coefficient update equations for the truncated Volterra filters given above. The input signal x(n) was generated by passing a zero-mean white Gaussian noise through a linear filter with the following impulse response:

$$h_n = \begin{cases} 0.25; & \text{for } n = 0, \\ 1.0; & \text{for } n = 1, \\ 0.25; & \text{for } n = 2, \\ 0.0; & \text{otherwise} \end{cases}$$

The variance of the input was chosen to be 1. The desired response was chosen such that the signal to noise ratio was 30dB. Step-sizes were chosen such that the RLS and LMS-type algorithms will have the same excess mean-squared error at steady-state.

We ran several simulations using this structure.

The same system was adapted using the RLS-type adaptive algorithm.

Here we chose the same environment as the LMS-type algorithm and also $\lambda = 0.995$.

The performance of the LMS- and RLS-type algorithms for the second-order Volterra series model is shown in figure 6-4. The top part represents the performance measure for linear coefficients

$$\|V_{Linear}[n]\| = 10\log \frac{\sum_{i=0}^{N-1} (\hat{h}_1[i;n] - h_1[i;n])^2}{\sum_{i=0}^{N-1} (h_1[i;n])^2} \tag{6.14}$$

and the bottom part represents the performance measure for quadratic coefficients

$$\|V_{Quadratic}[n]\| = 10\log \frac{\sum_{i=0}^{N-1}\sum_{j=0}^{N-1} (\hat{h}_2[i,j;n] - h_2[i,j;n])^2}{\sum_{i=0}^{N-1}\sum_{j=0}^{N-1} (h_2[i,j;n])^2} \tag{6.15}$$

These are the norm-tap error measures in dB.

We see that the RLS converges faster than the LMS-type algorithm for both coefficients. Also, we note that after about 9,000 iterations, the quadratic coefficients using LMS-type algorithm eventually converge to the same results as the RLS-type algorithm. For the linear coefficients, there is some bias after convergence between the LMS and RLS results.

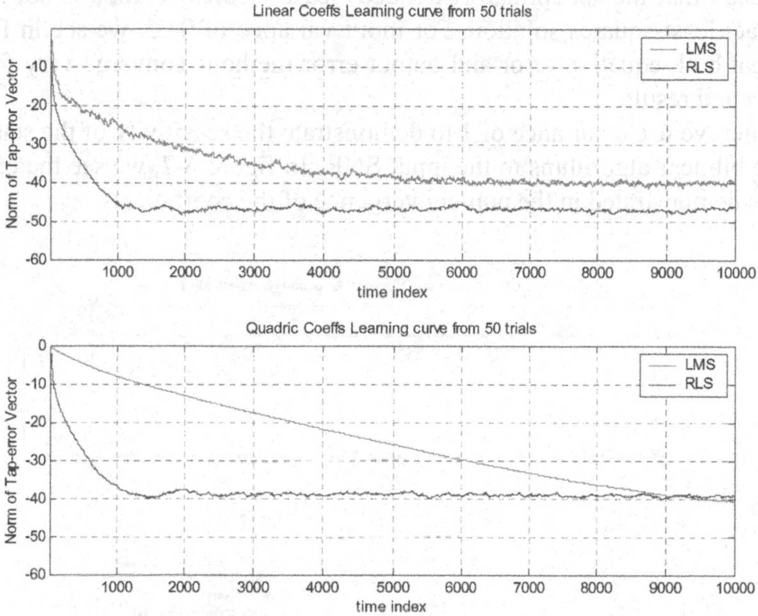

Figure 6-4. Performance of LMS and RLS-type adaptive filters for second-order Volterra series model—top graph is for linear coefficients, bottom graph is for quadratic coefficients.

Example 2:

We also simulated the LMS-type bilinear algorithm using the output error method and the equation error method separately. We use the following system:

$$y[n] = 0.9y[n-1] - 0.7y[n-2]x[n-1] + 0.5x[n-1]$$

which leads to the parameters: $a = 0.9, b = -0.7, c = 0.5$.

The input signal x(n) was a zero-mean white Gaussian noise with variance 0.05. In figure 6-5, we see that both equation error and output error methods converge very well to the desired result. However, we note that the estimates produced by the equation error method are indeed biased even for such low SNR.

Example 3:

We also simulated the RLS-type bilinear algorithm using the output error method and the equation error method but using the extended (suboptimal) least-squares method. We use the same bilinear system as in example 2.

Recall that the suboptimal (extended) least-squares solution is not really an exact least-squares solution. For input variance of 0.05, we see in figure 6-6 that both equation error and output error methods converge very fast to the desired result.

Later we use a variance of 1 to demonstrate the sensitivity of the stability of the bilinear algorithms to the input SNR. In figure 6-7, we see the effects of this demonstrated in the non-convergence of the coefficients.

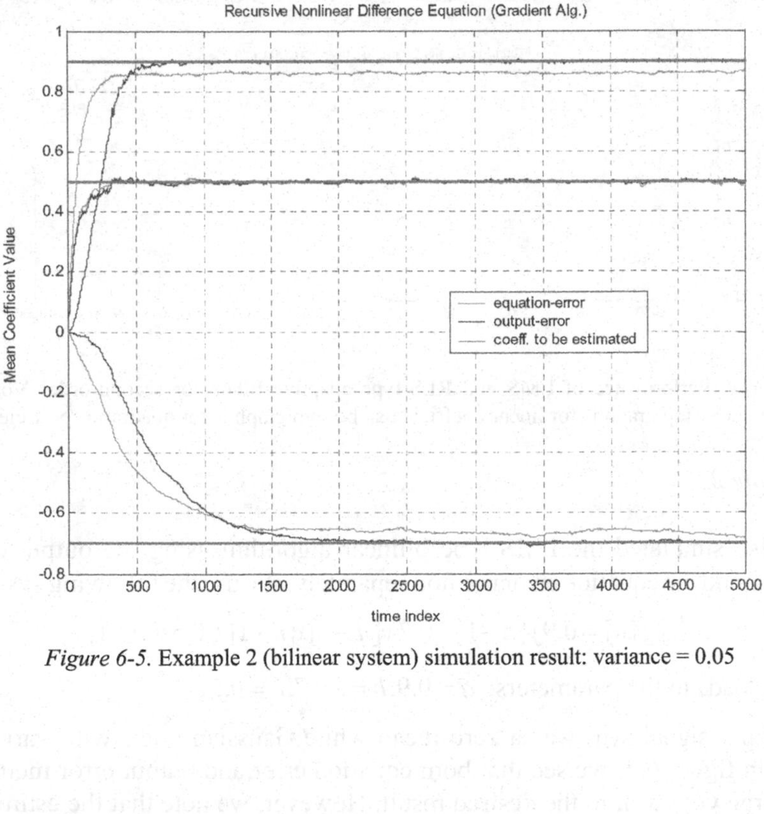

Figure 6-5. Example 2 (bilinear system) simulation result: variance = 0.05

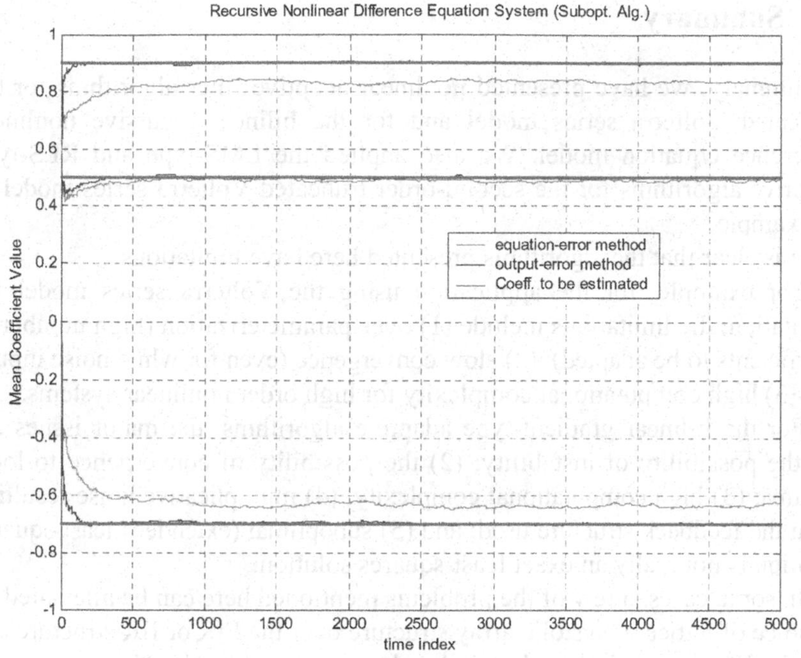

Figure 6-6. Example 3 (suboptimal bilinear system) simulation result: variance = 0.05

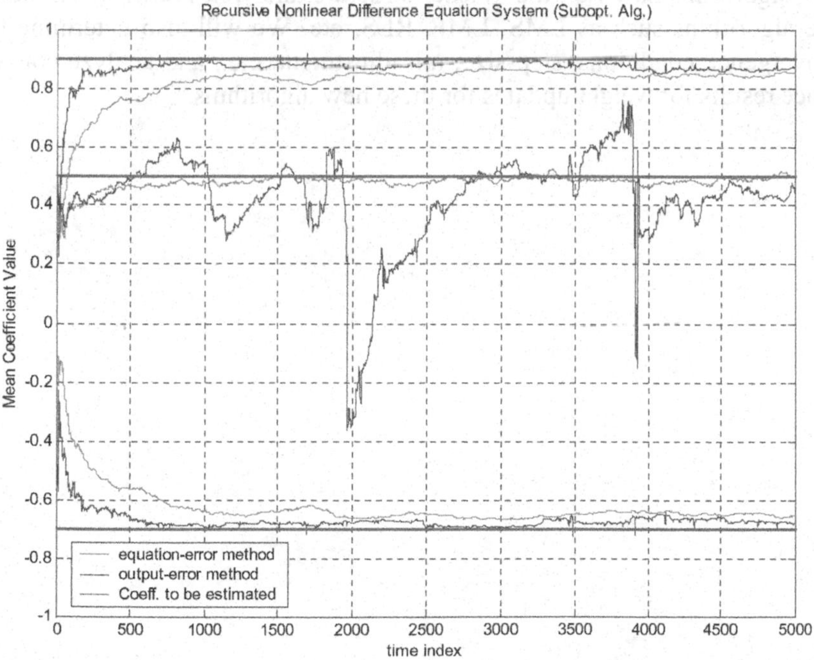

Figure 6-7. Example 3 (suboptimal bilinear system) simulation result: variance = 1.0

6.6 Summary

In summary, we have presented nonlinear adaptive filter algorithms for the truncated Volterra series model and for the bilinear recursive nonlinear difference equation model. We also applied the LMS-type and RLS-type adaptive algorithms for the second-order truncated Volterra series model as an example.

It is clear that the algorithms presented here have limitations.

For example, for the application using the Volterra series model for adaptation, the limitations include (1) over-parameterization (high number of coefficients to be adapted); (2) slow convergence (even for white noise input), and (3) high computational complexity for high order nonlinear systems.

For the bilinear gradient-type adaptive algorithms, the major issues are (1) the possibility of instability; (2) the possibility of convergence to local minima; (3) high computational complexity; (4) multiplicative noise resulting from the feedback structure used; and (5) suboptimal (extended) least-squares solution is not really an exact least-squares solution.

In some cases, a few of the problems mentioned here can be alleviated by a choice of lattice or systolic array structure over the FIR or IIR structures.

The Wiener model can be made adaptive (just like the Volterra model) and applied for system identification. In chapters 7 through 10, we will present new algorithms based on the Wiener structure and well known linear adaptive algorithms such as LMS, LMF, RLS, etc. We will also determine the convergence conditions, step-size, misadjustment, etc., and analyze convergence results for weight updates for these new algorithms.

Chapter 7

NONLINEAR ADAPTIVE SYSTEM IDENTIFICATION BASED ON WIENER MODELS (PART 1)

Second-order least-mean-square (LMS)-based approach

Introduction

In the previous chapter, we discussed several examples of adaptive nonlinear filtering using the truncated Volterra model. The first major paper to present this kind of adaptive algorithm was the paper by Koh and Powers (Koh 1983, 1985). Since then, there have been several other papers presenting similar algorithms. However, these algorithms have the following limitations: (1) They are over-parameterized (which means the number of coefficients is too high) even for low-order non-linear filters such as second-order. (2) The convergence rate is typically slow especially for stochastic-gradient-type algorithms. (3) They sometimes do not converge or lead to high mis-adjustment. (4) Their performance is not very good for non-white and/or non-Gaussian inputs.

In this chapter, given the background in chapter 3 (the relationship between the nonlinear Volterra and Wiener models), we are going to extend the developments of chapter 6 to a nonlinear LMS adaptive algorithm which is based on the discrete Wiener model. This nonlinear LMS adaptive method can be seen as an extension of the linear LMS algorithm. The merit of this approach is that it keeps most of the linear LMS properties but still has a reasonably good convergence rate. The performance analysis is also more tractable, which is seldom true for most Volterra model LMS algorithms (Ogunfunmi 2001, Chang 2003). Furthermore, the extension to a higher-order nonlinear Wiener model is straightforward, as we will show in chapter 8.

New algorithms are derived based on this Wiener structure and well known linear adaptive algorithms such as LMS, LMF, RLS, etc. Analysis of convergence-produced results for weight updates, convergence conditions, step-size, misadjustment, etc., for the new algorithms can be determined.

The Volterra series can be expanded by using the discrete Wiener model. This means that it is a proper subset of the Wiener class. However, for the Wiener model expansion, we need to know or find an orthonormal basis set for the expansion. The process of finding an orthonormal basis set for the Wiener model can be exemplified by using a tapped delay line for a Gaussian white input. The computational complexity of the Wiener model method is much less than for the Volterra model because the corresponding equivalent filter length is much less.

We consider developing an adaptive algorithm using the discrete non-linear Wiener model for second-order Volterra system identification application. First we have to perform a complete orthogonalization procedure on the truncated Volterra series. This allows us to use the LMS adaptive linear filtering algorithm to calculate all the coefficients recursively. This orthogonalization method is based on the nonlinear discrete Wiener model. It contains three sections: a single-input, multi-output linear with memory section; a multi-input, multi-output nonlinear no-memory section; and a multi-input, single-output amplification and summary section. For a white Gaussian noise input signal, the autocorrelation matrix of the adaptive filter input vector can be diagonalized, as is not the case when using the Volterra model. This dramatically reduces the eigenvalue spread and results in more rapid convergence. Also, the discrete nonlinear Wiener model adaptive system allows us to represent a complicated Volterra system with only a few coefficient terms. In general, it can also identify the nonlinear system without over-parameterization. The theoretical performance analysis of steady-state behavior is presented. Computer simulation examples are also included to verify the theory.

7.1 Second-Order System

7.1.1 Nonlinear LMS Adaptation Algorithm

To develop an adaptive algorithm based on the nonlinear Wiener model, consider the system identification application shown in figure 7-1:

The adaptive filter acts on the input sample x(n) to generate an estimate of the desired signal. For Gaussian white input and Gaussian white plant noise, it is possible to apply the nonlinear Wiener model in the adaptive plant block. This is because the $\widetilde{Q}$-polynomial has perfect orthogonality which allows us to perform the LMS algorithm with a reasonably good convergence rate. We apply the delay-line version discrete nonlinear Wiener model to the adaptive plant. The block diagram is shown in figure 7-2.

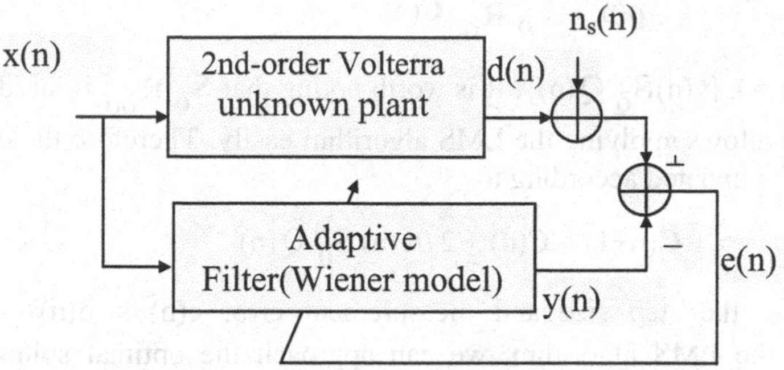

Figure 7-1. Second-order nonlinear system identification model

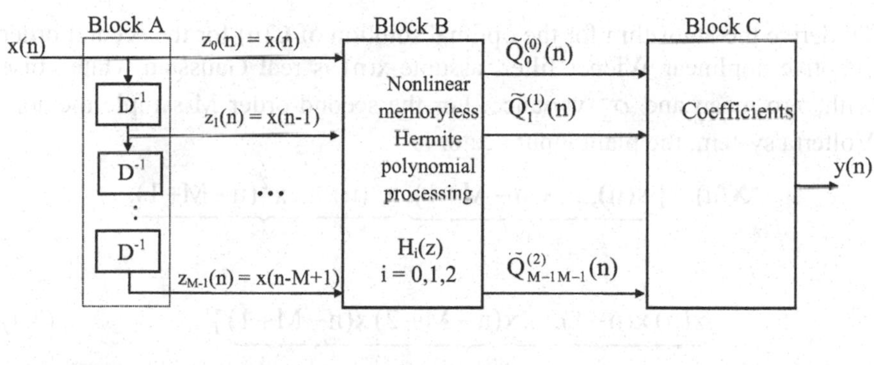

Figure 7-2. Delay line structure of a second-order nonlinear discrete Wiener model

To develop an adaptive algorithm, we need to write equation 3.88 in a matrix form

$$y(n) = \mathbf{S}_{\tilde{Q}}^{-1}\tilde{\mathbf{Q}}^T(n)\mathbf{C}(n) \tag{7.1}$$

which can be pre-calculated such that $\mathbf{S}_{\tilde{Q}}^{-1}\mathbf{R}_{\tilde{Q}\tilde{Q}}(n)\mathbf{S}_{\tilde{Q}}^{-1}$ becomes an identity matrix, where $\mathbf{R}_{\tilde{Q}\tilde{Q}} = E\{\tilde{\mathbf{Q}}(n)\tilde{\mathbf{Q}}^T(n)\}$. Note that $\mathbf{S}_{\tilde{Q}}^{-1}$ is not shown in equation 3.88. Pre-multiply equation 7.1 by $\mathbf{S}_{\tilde{Q}}^{-1}\tilde{\mathbf{Q}}(n)$ and take the expectation, using the properties of the $\widetilde{Q}$-polynomial in equation 3.85a and equation 3.85b. We get

$$\mathbf{p}(n) = \mathbf{S}_{\widetilde{Q}}^{-2} \mathbf{R}_{\widetilde{Q}\widetilde{Q}} \mathbf{C}(n) \tag{7.2}$$

where $\mathbf{p}(n) = E\{y(n)\mathbf{S}_{\widetilde{Q}}^{-1} \widetilde{\mathbf{Q}}(n)\}$. It is worth noting that $\mathbf{S}_{\widetilde{Q}}^{-2} \mathbf{R}_{\widetilde{Q}\widetilde{Q}}$ is an identity matrix that allows applying the LMS algorithm easily. Therefore the coefficients can be updated according to

$$\mathbf{C}(n+1) = \mathbf{C}(n) + 2\mu e(n)\, \mathbf{S}_{\widetilde{Q}}^{-1} \widetilde{\mathbf{Q}}(n) \tag{7.3}$$

where μ is the step size and measurement error $e(n) = d(n) - y(n)$. Following the LMS algorithm, we can approach the optimal solution of $\mathbf{C}(n)$, the minimum MSE (mean square error), the range of μ, time constant and misadjustment.

7.1.2 Second-Order System Performance Analysis

To derive the algorithm for the optimal solution of $\mathbf{C}(n)$ for the second-order adaptive nonlinear Wiener filter, assume x(n) is real Gaussian white noise with zero mean and σ_x^2 variance. For the second-order M-sample memory Volterra system, the plant input vector is

$$\mathbf{X}(n) = [\, \underbrace{x(n), \ldots, x(n-M+1)}_{M's}, \underbrace{x^2(n), \ldots, x^2(n-M+1)}_{M's},$$

$$\underbrace{x(n)\,x(n-1), \ldots, x(n-M+2)\,x(n-M+1)}_{M(M-1)/2's}\,]^T \tag{7.4}$$

The vector length of $\mathbf{X}(n)$ is $L=M(M+3)/2$. For the second-order M-sample memory truncated Volterra unknown plant, the output d(n) can be written as

$$d(n)=h_0+\sum_{k_1=0}^{M-1} h_1(k_1)\,x(n-k_1)+\ldots+\sum_{k_1=0}^{M-1}\sum_{k_2=0}^{M-1} h_2(k_1,k_2)\,x(n-k_1)\,x(n-k_2) \tag{7.5}$$

where $\{h_j(k_1, \ldots, k_j),\ 0 \le j \le 2\}$ is the set of second-order Volterra kernel coefficients. Assume that the kernels are symmetric, i.e., $h_2(k_1, k_2)$ is unchanged for the permutation of indices k_1 and k_2. One can think of each additional term of equation 7.5 as the one or two dimensional convolution. For ease of latter use, equation 7.5 can be expressed as a vector form which is

$$d(n) = \mathbf{C}^{*T}\mathbf{Q}(n) = \mathbf{C}^{*T}[1, \mathbf{X}^T(n)]^T = w_0^{\,*} + \mathbf{W}^{*T}\mathbf{X}(n) \tag{7.6}$$

where $\mathbf{C}^* = [w_0^*, w_1^*, ..., w_L^*]^T$ is the weight vector with length L+1, $\mathbf{Q}(n) = [1, \mathbf{X}^T(n)]^T$ and $\mathbf{W}^* = [w_1^*, ..., w_L^*]^T$. For simplicity and without loss of generality, we consider the delay-line version discrete nonlinear Wiener model shown in figure 7-2. In block A, there are M-1 delay elements. $H_0[z(n)] = z(n)$ and $H_1[z(n)] = z^2(n) - \sigma_x^2$ are applied in block B. The input vector of block C is

$$\widetilde{\mathbf{X}}(n) = [x(n), x(n-1), ..., x(n-M+1), x^2(n)-\sigma_x^2, x^2(n-M+1)-\sigma_x^2,$$

$$x(n)x(n-1), ..., x(n-M+1)x(n-M+2)]^T \tag{7.7}$$

The vector length $\widetilde{\mathbf{X}}(n)$ is the same as $\mathbf{X}(n)$. The output of the adapted plant y(n) is

$$y(n) = c_0 + \sum_{n_1=0}^{M-1} c_1(n_1)\widetilde{Q}_{n_1}^{(1)}(n) + \sum_{n_1=0}^{M-1}\sum_{n_2=0}^{M-1} c_2(n_1,n_2)\widetilde{Q}_{n_1n_2}^{(2)}(n) \tag{7.8}$$

where $\{c_j(n_1, .., n_j), 0 \le j \le 2\}$ is the set of second-order nonlinear Wiener kernel coefficients. We may express equation 7.8 as a vector form which is

$$y(n) = \mathbf{C}^T(n)\, \mathbf{S}_{\widetilde{Q}}^{-1}\, \widetilde{\mathbf{Q}}(n) = \mathbf{C}^T(n)\, \mathbf{S}_{\widetilde{Q}}^{-1}\, [1, \widetilde{\mathbf{X}}^T(n)]^T = w_0 + \mathbf{W}^T(n)\, \mathbf{S}_{\widetilde{X}}^{-1}\, \widetilde{\mathbf{X}}(n) \tag{7.9}$$

where $\mathbf{C}(n) = [w_0(n), \mathbf{W}^T(n)]^T$, $\mathbf{W}(n) = [w_1(n), ..., w_L(n)]^T$ and $\widetilde{\mathbf{Q}}(n) = [1, \widetilde{\mathbf{X}}^T(n)]^T$. $\mathbf{C}(n)$ is the coefficient vector in block C. If these coefficients are adapted to the proper values, then the model of thc adaptive plant will match the equivalent transfer function of the unknown plant. The error signal can be written

$$e(n) = d(n) - y(n) + n_s(n)$$

$$= \mathbf{C}^{*T}\mathbf{Q}(n) - \mathbf{C}^T(n)\, \mathbf{S}_{\widetilde{Q}}^{-1}\, \widetilde{\mathbf{Q}}(n) + n_s(n) \tag{7.10a}$$

$$= w_0^* - w_0(n) + \mathbf{W}^{*T}\mathbf{X}(n) - \mathbf{W}^T(n)\, \mathbf{S}_{\widetilde{X}}^{-1}\, \widetilde{\mathbf{X}}(n) + n_s(n) \tag{7.10b}$$

Then if we expand equation 7.10b and take expectation, we find the mean square error

$$\xi(n) = E\{e^2(n)\}$$

$$\begin{aligned}\xi(n) = {} & E\{[n_s(n)+w_0^{*2} - w_0^2(n)]^2\} + \mathbf{W}^{*T}\mathbf{R}_{XX}\mathbf{W}^* - \mathbf{W}^T(n)\mathbf{S}_{\widetilde{X}}^{-1}\, \mathbf{R}_{\widetilde{X}\widetilde{X}}\, \mathbf{S}_{\widetilde{X}}^{-1}\, \mathbf{W}(n) \\ & + 2E\{n_s(n)\mathbf{X}^T(n)\}\mathbf{W}^* \\ & + 2w_0^* E\{\mathbf{X}^T(n)\}\mathbf{W}^* - 2w_0(n) - E\{\mathbf{X}^T(n)\}\mathbf{W}^* - 2E\{n_s(n)\, \widetilde{\mathbf{X}}^T(n)\}\, \mathbf{S}_{\widetilde{X}}^{-1}\mathbf{W}(n) \\ & - 2w_0^* E\{\widetilde{\mathbf{X}}^T(n)\}\mathbf{S}_{\widetilde{X}}^{-1}\mathbf{W}(n) + 2w_0(n)E\{\widetilde{\mathbf{X}}^T(n)\}\mathbf{W}(n) - 2\mathbf{W}^T(n)\mathbf{S}_{\widetilde{X}}^{-1}\, \mathbf{R}_{\widetilde{X}X}\mathbf{W}^*\end{aligned} \tag{7.11}$$

where $\mathbf{R}_{\mathbf{XX}}= \{\mathbf{X}(n)\mathbf{X}^T(n)\}$, $\mathbf{R}_{\tilde{\mathbf{X}}\tilde{\mathbf{X}}}= E\{\tilde{\mathbf{X}}(n)\tilde{\mathbf{X}}^T(n)\}$ and $\mathbf{R}_{\tilde{\mathbf{X}}\mathbf{X}}= E\{\tilde{\mathbf{X}}(n)\mathbf{X}^T(n)\}$. Assume that the plant noise $n_s(n)$ is uncorrelated with the input signal $\mathbf{X}(n)$, therefore $E\{n_s(n)\mathbf{X}^T(n)\} = \mathbf{0}$, where $\mathbf{0}$ is a column vector. We also note that $E\{\tilde{\mathbf{X}}^T(n)\} = \mathbf{0}$ because of the zero mean property of $\tilde{Q}$-polynomial. In steady state, $[w_o(n)\ \mathbf{W}^T(n)]^T$ does not change with time. Then equation 7.11 can be simplified as

$$\xi(n) = \sigma^2_{n_s} + w_0^{*2} - w_0^2 - 2w_0^* w_0 + 2w_0^* E\{\mathbf{X}^T(n)\}\mathbf{W}^* - 2w_0(n)E\{\mathbf{X}^T(n)\}\mathbf{W}^*$$

$$+ \mathbf{W}^{*T}\mathbf{R}_{\mathbf{XX}}\mathbf{W}^* + \mathbf{W}^T\mathbf{S}_{\tilde{\mathbf{X}}}^{-1}\mathbf{R}_{\tilde{\mathbf{X}}\tilde{\mathbf{X}}}\mathbf{S}_{\tilde{\mathbf{X}}}^{-1}\mathbf{W} - 2\mathbf{W}^T\mathbf{S}_{\tilde{\mathbf{X}}}^{-1}\mathbf{R}_{\tilde{\mathbf{X}}\mathbf{X}}\mathbf{W}^* \qquad (7.12)$$

where $\sigma^2_{n_s} = E\{n_s^2(n)\}$. To minimize equation 7.12, we need to take the derivative with respect to $\mathbf{C}= [w_0, \mathbf{W}^T]^T$, which is

$$\frac{\partial\xi(n)}{\partial w_0} = 2w_0^* - 2w_0 - 2E\{\mathbf{X}^T(n)\}\mathbf{W}^* = 0 \qquad (7.13a)$$

$$\frac{\partial\xi(n)}{\partial \mathbf{W}} = 2\mathbf{S}_{\tilde{\mathbf{X}}}^{-1}\mathbf{R}_{\tilde{\mathbf{X}}\tilde{\mathbf{X}}}\mathbf{S}_{\tilde{\mathbf{X}}}^{-1}\mathbf{W} - 2\mathbf{S}_{\tilde{\mathbf{X}}}^{-1}\mathbf{R}_{\tilde{\mathbf{X}}\mathbf{X}}\mathbf{W}^* = \mathbf{0} \qquad (7.13b)$$

Note that $\mathbf{R}_{\tilde{\mathbf{X}}\tilde{\mathbf{X}}}= \mathbf{R}_{\tilde{\mathbf{X}}\mathbf{X}}$ (see appendix 7A), then the optimal solutions can be obtained as

$$w_{0/optm}= w_0^* + E\{\mathbf{X}^T(n)\}\mathbf{W}^* \qquad (7.14a)$$

$$\mathbf{W}_{optm}= \mathbf{S}_{\tilde{\mathbf{X}}}\mathbf{W}^* \qquad (7.14b)$$

In order to obtain the MMSE solution, we assume that the filter coefficients have converged. Substituting equation 7.14a and equation 7.14b in equation 7.10, the minimum mean square error is

$$\xi_{min}= E\{n_s^2(n)\} = \sigma^2_{n_s} \qquad (7.15)$$

It is interesting to note that the unknown plant is a Volterra model and the adapted plant is a nonlinear Wiener model, both of which are not linear systems. But from the derivations, we can see that equation 7.13 and equation 7.14 have almost the same forms as the LMS algorithm for linear systems.

To derive the step size range, we need to consider the instantaneous version of equation 7.12. Define the error power as

$$\varepsilon(\mathrm{n}) = e^2(n)$$

$$\varepsilon(\mathrm{n}) = [\mathrm{n_s(n)} + \mathrm{w}_0^* - \mathrm{w_0(n)}]^2 + \mathbf{W}^{*\mathrm{T}}\mathbf{X}(\mathrm{n})\mathbf{X}^{\mathrm{T}}(\mathrm{n})\,\mathbf{W}^* + \mathbf{W}^{\mathrm{T}}(\mathrm{n})\,\mathbf{S}_{\widetilde{\mathbf{X}}}^{-1}\,\widetilde{\mathbf{X}}(\mathrm{n})\,\widetilde{\mathbf{X}}^{\mathrm{T}}(\mathrm{n})\,\mathbf{S}_{\widetilde{\mathbf{X}}}^{-1}\,\mathbf{W}(\mathrm{n})$$

$$+2\mathrm{n_s(n)}\,\mathbf{X}^{\mathrm{T}}(\mathrm{n})\mathbf{W}^* + 2\mathrm{w}_0^*\,\mathbf{X}^{\mathrm{T}}(\mathrm{n})\mathbf{W}^* - 2\mathrm{w_0(n)}\mathbf{X}^{\mathrm{T}}(\mathrm{n})\mathbf{W}^* - 2\mathrm{n_s(n)}\,\widetilde{\mathbf{X}}^{\mathrm{T}}(\mathrm{n})\,\mathbf{S}_{\widetilde{\mathbf{X}}}^{-1}\,\mathbf{W}(\mathrm{n})$$

$$-2\mathrm{w}_0^*\widetilde{\mathbf{X}}^{\mathrm{T}}(\mathrm{n})\,\mathbf{S}_{\widetilde{\mathbf{X}}}^{-1}\,\mathbf{W}(\mathrm{n}) - 2\mathrm{w_0(n)}\,\widetilde{\mathbf{X}}^{\mathrm{T}}(\mathrm{n})\,\mathbf{S}_{\widetilde{\mathbf{X}}}^{-1}\,\mathbf{W}(\mathrm{n}) - 2\mathbf{W}^{\mathrm{T}}(\mathrm{n})\,\mathbf{S}_{\widetilde{\mathbf{X}}}^{-1}\,\widetilde{\mathbf{X}}(\mathrm{n})\,\mathbf{X}^{\mathrm{T}}(\mathrm{n})\mathbf{W}^* \qquad (7.16)$$

Taking the derivative of equation 7.16, we have

$$\frac{\partial\varepsilon(n)}{\partial \mathrm{w}_0} = -2[\mathrm{n_s(n)} + \mathrm{w}_0^* + \mathbf{X}^{\mathrm{T}}(\mathrm{n})\mathbf{W}^* - \mathrm{w_0(n)} - \widetilde{\mathbf{X}}^{\mathrm{T}}(\mathrm{n})\mathbf{S}_{\widetilde{\mathbf{X}}}^{-1}\mathbf{W}(\mathrm{n})\,] = -2\mathrm{e(n)} \qquad (7.17\mathrm{a})$$

$$\frac{\partial\varepsilon(\mathrm{n})}{\partial \mathbf{W}} = -2\mathbf{S}_{\widetilde{\mathbf{X}}}^{-1}\,\widetilde{\mathbf{X}}^{\mathrm{T}}(\mathrm{n})[\,\mathrm{n_s(n)} + \mathrm{w}_0^* + \mathbf{X}^{\mathrm{T}}(\mathrm{n})\mathbf{W}^* - \mathrm{w_0(n)} - \widetilde{\mathbf{X}}^{\mathbf{T}}(\mathrm{n})\mathbf{S}_{\widetilde{\mathbf{X}}}^{-1}\mathbf{W}(\mathrm{n})]$$

$$= -2\,\mathbf{S}_{\widetilde{\mathbf{X}}}^{-1}\,\widetilde{\mathbf{X}}^{\mathrm{T}}(\mathrm{n})\mathrm{e(n)} \qquad (7.17\mathrm{b})$$

Based on the steepest descent method, the weight updates for equation 7.17a and equation 7.17b are

$$\mathbf{C}(\mathrm{n+1}) = \mathbf{C}(\mathrm{n}) - \mu\nabla\varepsilon(\mathrm{n})\big|_{\mathbf{C}=\mathbf{C}(\mathrm{n})} = \begin{bmatrix} \mathrm{w_0(n)} - \mu\,\partial\varepsilon(\mathrm{n})/\partial\,\mathrm{w}_0 \\ \mathbf{W}(\mathrm{n}) - \mu\,\partial\varepsilon(\mathrm{n})/\partial\mathbf{W} \end{bmatrix}$$

$$= \begin{bmatrix} \mathrm{w_0(n)} \\ \mathbf{W}(\mathrm{n}) \end{bmatrix} - \mu \begin{bmatrix} -2\,\mathrm{e(n)} \\ -2\,\mathrm{e(n)}\mathbf{S}_{\widetilde{\mathbf{X}}}^{-1}\widetilde{\mathbf{X}}(\mathrm{n}) \end{bmatrix}$$

$$= \mathbf{C}(\mathrm{n}) + 2\mu\mathrm{e(n)} \begin{bmatrix} 1 & \mathbf{0}^{\mathrm{T}} \\ \mathbf{0} & \mathbf{S}_{\widetilde{\mathbf{X}}}^{-1} \end{bmatrix} \begin{bmatrix} 1 \\ \widetilde{\mathbf{X}}(\mathrm{n}) \end{bmatrix}$$

$$\mathbf{C}(\mathrm{n+1}) = \mathbf{C}(\mathrm{n}) + 2\mu\mathrm{e(n)}\mathbf{S}_{\widetilde{\mathbf{Q}}}^{-1}\,\widetilde{\mathbf{Q}}(\mathrm{n}) \qquad (7.18)$$

where ∇ means true gradient operator, $\mathbf{S}_{\widetilde{\mathbf{Q}}}^{-1} = \begin{bmatrix} 1 & \mathbf{0}^{\mathrm{T}} \\ \mathbf{0} & \mathbf{S}_{\widetilde{\mathbf{X}}}^{-1} \end{bmatrix}$ and $\mathbf{0}$ is a column vector. This is similar to but different from the LMS algorithm for linear systems. Substituting equation 7.10a in equation 7.18 and taking expectation on both sides, we obtain

$$E\{\mathbf{C}(n+1)\} = (\mathbf{I}-2\mu\mathbf{S}_{\tilde{Q}}^{-1}\mathbf{R}_{\tilde{Q}\tilde{Q}}\mathbf{S}_{\tilde{Q}}^{-1})E\{\mathbf{C}(n)\} + 2\mu\mathbf{S}_{\tilde{Q}}^{-1}\mathbf{R}_{\tilde{Q}Q}\mathbf{C}^{*}+2\mu\begin{bmatrix} E\{\mathbf{X}^{T}(n)\}\mathbf{W}^{*} \\ \mathbf{0} \end{bmatrix} \quad (7.19)$$

where **I** is an identity matrix. Define the weight error vector as

$$\mathbf{V}(n) = \mathbf{S}_{\tilde{Q}}^{-1}(\mathbf{C}(n) - \mathbf{C}_{optm}) \quad (7.20)$$

By the definition of equation 7.20, replacing $\mathbf{C}(n)$ with $\mathbf{S}_{\tilde{Q}}\mathbf{V}(n) + \mathbf{C}_{optm}$ in equation 7.19, and using the fact of $\mathbf{R}_{\tilde{Q}\tilde{Q}} = \mathbf{R}_{\tilde{Q}Q}$ yields the recursive version of $\mathbf{V}(n)$:

$$E\{\mathbf{V}(n+1)\} = (\mathbf{I}-2\mu\mathbf{S}_{\tilde{Q}}^{-2}\mathbf{R}_{\tilde{Q}\tilde{Q}})E\{\mathbf{V}(n)\} \quad (7.21)$$

Note that $\mathbf{R}_{\tilde{Q}\tilde{Q}} = \tilde{\mathbf{P}}\tilde{\mathbf{D}}\tilde{\mathbf{P}}^{-1}$ where $\tilde{\mathbf{P}}$ and $\tilde{\mathbf{D}}$ are eigenvector square matrix and eigenvalue square matrix corresponding to $\mathbf{R}_{\tilde{Q}\tilde{Q}}$ respectively (Widrow 1985). Substituting $\tilde{\mathbf{P}}^{-1}\mathbf{V}(n)$ with $\mathbf{V}'(n)$ in equation 7.21, and multiplying by $\tilde{\mathbf{P}}^{-1}$ on both sides, we obtain

$$E\{\mathbf{V}'(n)\} = (I-2\mu\mathbf{S}_{\tilde{Q}}^{-2}\tilde{\mathbf{D}})^{n}\mathbf{V}'(0) \quad (7.22)$$

Therefore, for convergence, the range of step size should be

$$0<\mu<\frac{1}{\lambda_{max}} \quad \text{or} \quad 0<\mu<\frac{1}{tr[\mathbf{S}_{\tilde{Q}}^{-2}\mathbf{R}_{\tilde{Q}\tilde{Q}}]}<\frac{1}{\lambda_{max}} \quad (7.23)$$

where λ_{max} is the maximum eigenvalue of $\mathbf{S}_{\tilde{Q}}^{-2}\mathbf{R}_{\tilde{Q}\tilde{Q}}$. Because $\mathbf{S}_{\tilde{Q}}^{-2}\mathbf{R}_{\tilde{Q}\tilde{Q}}$ is an identity matrix, this means $\lambda_{max}=1$, then equation 7.23 becomes

$$0<\mu<1 \quad \text{or} \quad 0<\mu<\frac{1}{tr[\mathbf{S}_{\tilde{Q}}^{-2}\mathbf{R}_{\tilde{Q}\tilde{Q}}]}<1 \quad (7.24)$$

To derive the misadjustment, we need to consider the steady state condition. The steady state of equation 7.12 can be expressed as

$$\begin{aligned}\hat{\xi}(n) =\ & E\{[n_s(n)+w_0^{*2}-\hat{w}_0(n)]^2\}+\mathbf{W}^{*T}\mathbf{R}_{XX}\mathbf{W}^{*}-\hat{\mathbf{W}}_0^{T}(n)\mathbf{S}_{\tilde{X}}^{-1}\mathbf{R}_{\tilde{X}\tilde{X}}\mathbf{S}_{\tilde{X}}^{-1}\hat{\mathbf{W}}_0^{T}(n)\\ & +2w_0^{*}E\{\mathbf{X}^{T}(n)\}\mathbf{W}^{*}-2\hat{w}_0(n)E\{\mathbf{X}^{T}(n)\}\mathbf{W}^{*}-2\hat{\mathbf{W}}_0^{T}(n)\mathbf{S}_{\tilde{X}}^{-1}\mathbf{R}_{\tilde{X}X}\mathbf{W}^{*}\\ =\ & \sigma_{n_s}^2+w_0^{*2}+\hat{w}_0^2(n)-2w_0^{*}\hat{w}_0^2(n)+2w_0^{*}E\{\mathbf{X}^{T}(n)\}\mathbf{W}^{*}\\ & -2\hat{w}_0(n)E\{\mathbf{X}^{T}(n)\}\mathbf{W}^{*}\\ & +\mathbf{W}^{*T}\mathbf{R}_{XX}\mathbf{W}^{*}+\hat{\mathbf{W}}^{T}(n)\mathbf{S}_{\tilde{X}}^{-1}\mathbf{R}_{\tilde{X}\tilde{X}}\mathbf{S}_{\tilde{X}}^{-1}-2\hat{\mathbf{W}}^{T}(n)\mathbf{S}_{\tilde{X}}^{-1}\mathbf{R}_{\tilde{X}X}\mathbf{W}^{*} \quad (7.25)\end{aligned}$$

where the header ^ means steady state. In steady state, we can assume that $\hat{\mathbf{C}}^T(n) \approx \mathbf{C}_{optm}$, which implies that $\hat{w}_0(n) \approx w_0^* + E\{\mathbf{X}^T(n)\}\mathbf{W}^*$ and $\hat{\mathbf{W}}_0^T(n) \approx \mathbf{S}_{\tilde{X}}\mathbf{W}^*$. Therefore, equation 7.25 can be expanded and simplified as:

$$\hat{\xi}(n) = \xi_{min} - \mathbf{W}^{*T}\mathbf{M}_x\mathbf{M}_x^T\mathbf{W}^* + \mathbf{W}^{*T}\mathbf{R}_{XX}\mathbf{W}^* + \hat{\mathbf{W}}^T(n)\mathbf{S}_{\tilde{X}}^{-1}\mathbf{R}_{\tilde{X}\tilde{X}}\mathbf{S}_{\tilde{X}}^{-1} - 2\hat{\mathbf{W}}^T(n)\mathbf{S}_{\tilde{X}}^{-1}\mathbf{R}_{\tilde{X}X}\mathbf{W}^* \tag{7.26}$$

where $\mathbf{M}_x = E\{\mathbf{X}(n)\}$. Using the relationships $\mathbf{R}_{XX} = \mathbf{R}_{\tilde{X}\tilde{X}} + \mathbf{M}_x\mathbf{M}_x^T$ (see appendix 7B) and $\mathbf{R}_{\tilde{X}\tilde{X}} = \mathbf{R}_{\tilde{X}X}$ (see appendix 7A), with proper manipulation, equation 7.26 can be approximated as

$$\begin{aligned}\hat{\xi}(n) &\approx \xi_{min} - \mathbf{W}^*\mathbf{M}_x\mathbf{M}_x^T\mathbf{W}^* + \mathbf{W}^{*T}\mathbf{M}_x\mathbf{M}_x^T\mathbf{W} \\ &\quad + [\mathbf{W}^{*T} - \mathbf{S}_{\tilde{X}}^{-1}\hat{\mathbf{W}}^T(n)]\mathbf{R}_{\tilde{X}\tilde{X}}[\mathbf{W}^* - \mathbf{S}_{\tilde{X}}^{-1}\hat{\mathbf{W}}(n)] \\ &\approx \xi_{min} + [\mathbf{C}_{optm} - \hat{\mathbf{C}}(n)]^T\mathbf{S}_{\tilde{Q}}^{-1}\mathbf{R}_{\tilde{X}\tilde{X}}\mathbf{S}_{\tilde{Q}}^{-1}[\mathbf{C}_{optm} - \hat{\mathbf{C}}(n)] \\ &\approx \xi_{min} + \hat{\mathbf{V}}^T(n)\mathbf{R}_{\tilde{Q}\tilde{Q}}\hat{\mathbf{V}}(n)\end{aligned} \tag{7.27}$$

where $\hat{\mathbf{V}}(n) = \mathbf{S}_{\tilde{Q}}^{-1}(\hat{\mathbf{C}}(n) - \mathbf{C}_{optm})$. We define the excess mean-square error as

$$\begin{aligned}\text{excess MSE} = E\{\hat{\xi}(n) - \xi_{min}\} &= E\{\hat{\mathbf{V}}^T(n)\mathbf{R}_{\tilde{Q}\tilde{Q}}\hat{\mathbf{V}}(n)\} \\ &= E\{\hat{\mathbf{V}}^T(n)\tilde{\mathbf{P}}^T\tilde{\mathbf{D}}\tilde{\mathbf{P}}\hat{\mathbf{V}}(n)\} \\ &= E\{\hat{\mathbf{V}}'^T(n)\tilde{\mathbf{D}}\hat{\mathbf{V}}'(n)\}\end{aligned} \tag{7.28}$$

Recall that $\tilde{\mathbf{P}}$ and $\tilde{\mathbf{D}}$ are, respectively, the eigenvector square matrix and the eigenvalue square matrix corresponding to $\mathbf{R}_{\tilde{Q}\tilde{Q}}$ and $\hat{\mathbf{V}}'(n) = \tilde{\mathbf{P}}^T\hat{\mathbf{V}}(n)$. Now consider the steady state error gradient

$$\hat{\nabla}\varepsilon(n) = \nabla\varepsilon(n) + \mathbf{N}_s(n) \tag{7.29}$$

where $\mathbf{N}_s(n)$ is defined as a gradient estimation noise vector (Widrow 1985). At steady state, $\nabla\varepsilon(n) \approx 0$, then

$$\hat{\nabla}\varepsilon(n) \approx \mathbf{N}_s(n) = -2e(n)\mathbf{S}_{\tilde{Q}}^{-1}\tilde{\mathbf{Q}}(n) \tag{7.30}$$

To find the covariance $\mathrm{cov}\{\hat{\mathbf{V}}'(n)\} = E\{\hat{\mathbf{V}}'(n)\hat{\mathbf{V}}'^{T}(n)\}$, we should first assume e(n) and $\widetilde{\mathbf{Q}}(n)$ are independent, then the covariance of $\mathbf{N}_s(n)$ can be approximated as

$$\mathrm{cov}[\mathbf{N}_s(n)] \approx 4E\{e^2(n)\}\, \mathbf{S}_{\widetilde{Q}}^{-1}\, E\{\widetilde{\mathbf{Q}}(n)\widetilde{\mathbf{Q}}^{T}(n)\}\,\mathbf{S}_{\widetilde{Q}}^{-1} = 4\xi_{min}\, \mathbf{S}_{\widetilde{Q}}^{-1}\mathbf{R}_{\widetilde{Q}\widetilde{Q}}\mathbf{S}_{\widetilde{Q}}^{-1} \quad (7.31)$$

For the second step, we define $\mathbf{N}'_s(n) = \widetilde{\mathbf{P}}^{-1}\mathbf{N}_s(n)$ and we find the covariance of $\mathbf{N}'_s$ which is

$$\mathrm{cov}[\mathbf{N}'_s(n)] = \mathrm{cov}[\widetilde{\mathbf{P}}^{-1}\mathbf{N}_s(n)] = \widetilde{\mathbf{P}}^{-1}E\{\mathbf{N}_s(n)\mathbf{N}_s^{T}(n)\}\,\widetilde{\mathbf{P}} = 4\xi_{min}\,\mathbf{S}_{\widetilde{Q}}^{-1}\widetilde{\mathbf{D}}\mathbf{S}_{\widetilde{Q}}^{-1} \quad (7.32)$$

For the third step, we need to find the covariance of $\mathbf{V}'(n)$.

To develop the LMS-type algorithm, we proceed as follows. From the steepest-descent method, we note that

$$\hat{\mathbf{C}}(n+1) = \hat{\mathbf{C}}(n) - \mu\hat{\nabla}\xi(n) \quad (7.33)$$

In equation 7.33, we note that $\hat{\mathbf{C}}(n) \approx \mathbf{S}_{\widetilde{Q}}\mathbf{V}(n) + \mathbf{C}_{optm}$ and

$\hat{\nabla}\xi(n) = 2\mathbf{S}_{\widetilde{Q}}^{-1}\mathbf{R}_{\widetilde{Q}\widetilde{Q}}\hat{\mathbf{V}}(n) + \mathbf{N}_s(n)$, therefore, equation 7.33 becomes

$$\hat{\mathbf{V}}(n+1) = (\mathbf{I} - 2\xi\,\mathbf{S}_{\widetilde{Q}}^{-2}\widetilde{\mathbf{D}})\hat{\mathbf{V}}(n) - \mu\,\mathbf{S}_{\widetilde{Q}}^{-1}\mathbf{N}_s(n) \quad (7.34)$$

Recall that $\hat{\mathbf{V}}'(n) = \widetilde{\mathbf{P}}^{-1}\hat{\mathbf{V}}(n)$ or $\hat{\mathbf{V}}(n) = \widetilde{\mathbf{P}}^{T}\hat{\mathbf{V}}'(n)$. Thus,

$$\hat{\mathbf{V}}'(n+1) = (\mathbf{I} - 2\mu\mathbf{S}_{\widetilde{Q}}^{-2}\widetilde{\mathbf{D}})\,\hat{\mathbf{V}}'(n) - \mu\mathbf{S}_{\widetilde{Q}}^{-1}\mathbf{N}'_s(n) \quad (7.35)$$

where $\mathbf{N}'_s(n) = \widetilde{\mathbf{P}}^{-1}\mathbf{N}_s(n)$. Then by equation 7.35, we can find the covariance of $\hat{\mathbf{V}}'(n)$ which is

$$\begin{aligned}\mathrm{cov}\{\hat{\mathbf{V}}'(n)\} &= E\{[\mathbf{I} - 2\mu\mathbf{S}_{\widetilde{Q}}^{-2}\widetilde{\mathbf{D}})\hat{\mathbf{V}}'(n) - \mu\mathbf{S}_{\widetilde{Q}}^{-1}\mathbf{N}'_s(n)][(\mathbf{I} - 2\mu\mathbf{S}_{\widetilde{Q}}^{-2}\widetilde{\mathbf{D}})\\ &\quad \hat{\mathbf{V}}'(n) - \mu\mathbf{S}_{\widetilde{Q}}^{-1}\mathbf{N}'_s(n)]^{T}\}\\ &= (\mathbf{I} - 2\mu\mathbf{S}_{\widetilde{Q}}^{-2}\widetilde{\mathbf{D}})\,E\{\hat{\mathbf{V}}'(n-1)\hat{\mathbf{V}}'^{T}(n-1)\}\,(\mathbf{I} - 2\mu\mathbf{S}_{\widetilde{Q}}^{-2}\widetilde{\mathbf{D}})^{T}\\ &\quad + \mu^2\mathbf{S}_{\widetilde{Q}}^{-1}E\{\mathbf{N}'_s(n-1)\mathbf{N}'^{T}_s(n-1)\}\,\mathbf{S}_{\widetilde{Q}}^{-1}\end{aligned} \quad (7.36)$$

We arrived at equation 7.34 because we assumed $\hat{\mathbf{V}}'(n)$ and $\mathbf{N}'_s(n)$ are independent, which means that $E\{\hat{\mathbf{V}}'(n)\mathbf{N}'^{T}_s(n)\} = 0$. Rearranging equation 7.34 by collecting all the $\mathrm{cov}\{\widetilde{\mathbf{V}}'(n)\}$ terms to the left hand side, we have

$$\mathrm{cov}\{\hat{\mathbf{V}}'(\mathrm{n})\} = \frac{\mu}{4(\mathbf{S}_{\tilde{\mathrm{Q}}}^{-2}\widetilde{\mathbf{D}} - \mu\mathbf{S}_{\tilde{\mathrm{Q}}}^{-4}\widetilde{\mathbf{D}}^2)}\ \mathrm{cov}\{\mathbf{N}'_{\mathrm{s}}(\mathrm{n})\}\ \mathbf{S}_{\tilde{\mathrm{Q}}}^{-1} \qquad (7.37)$$

Use $\mu\mathbf{S}_{\tilde{\mathrm{Q}}}^{-4}\widetilde{\mathbf{D}}^2 << \mathbf{I}$, and equation 7.32, equation 7.37 can be simplified as

$$\mathrm{cov}\{\hat{\mathbf{V}}'(\mathrm{n})\} \approx \mu\ \xi_{\min}\ \mathbf{S}_{\tilde{\mathrm{Q}}}^{-2} \qquad (7.38)$$

Finally, we obtain the excess mean square error as

$$\text{excess MSE} = \mathrm{E}\{\hat{\mathbf{V}}'^{\mathrm{T}}(\mathrm{n})\widetilde{\mathbf{D}}\,\hat{\mathbf{V}}'(\mathrm{n})\} \approx \mu\ \xi_{\min}\ \mathrm{tr}[\mathbf{S}_{\tilde{\mathrm{Q}}}^{-2}\mathbf{R}_{\tilde{\mathrm{Q}}\tilde{\mathrm{Q}}}] \qquad (7.39)$$

The misadjustment is defined as (Be 1985)

$$\mathrm{MISADJ} = \frac{\text{excess MSE}}{\xi_{\min}} \approx \mu\mathrm{tr}[\mathbf{S}_{\tilde{\mathrm{Q}}}^{-2}\mathbf{R}_{\tilde{\mathrm{Q}}\tilde{\mathrm{Q}}}] \qquad (7.40)$$

For the time-constant analysis, let us rewrite equation 7.22 in scalar form

$$\mathrm{v}'(\mathrm{n}) = (1 - 2\mu\lambda_{\mathrm{n}})^{\mathrm{n}}\ \mathrm{v}'(0) \qquad (7.41)$$

where λ_{n} is the eigenvalue of $\mathbf{S}_{\tilde{\mathrm{Q}}}^{-2}\mathbf{D}$. The term $(1-2\mu\lambda_{\mathrm{n}})^{\mathrm{n}}$ can be approximated by $(\mathrm{e}^{1/\tau_{\mathrm{n}}})^{\mathrm{n}}$.

If τ_{n} is large, then (Haykin 1996)

$$1 - 2\mu\lambda_{\mathrm{n}} \approx \mathrm{e}^{1/\tau_{\mathrm{n}}} \approx 1 - \frac{1}{\tau_n} \qquad (7.42)$$

The time-constant can be found as

$$\tau_{\mathrm{n}} \approx \frac{1}{2\mu\lambda_{\mathrm{n}}} \qquad (7.43)$$

Note that λ_{n} is the eigenvalue of an identity matrix; then 7.41 can simply be written as

$$\tau_{\mathrm{n}} \approx \frac{1}{2\mu} \qquad (7.44)$$

A count of the arithmetic operations involved in the implementation of the algorithm listed in table 7.1 shows that it requires $2\mathrm{M}^2+5\mathrm{M}+3$ multiplications per iteration. Our approach therefore has $\mathrm{O}(\mathrm{M}^2)$ computational complexity.

The extension of our algorithm and performance analysis to identify third-order Volterra systems will be presented in the following chapter.

Table 7.1. Computational complexity of the second-order nonlinear Wiener LMS adaptive filter

Initialization:	$\mathbf{C}(n) = \mathbf{0}$		
	Pre-calculate $\mathbf{S}_{\tilde{Q}}^{-1}$		
	$\mu' = 2\mu$		
	Relation	Dimension	Multiplication Count
Block A	$\mathbf{Z}(n) = [x(n), x(n-1), \ldots, x(m-M+1)]^T$	$M\times 1$	0
Block B	$\tilde{\mathbf{Q}}(n) = [1, \tilde{\mathbf{X}}^T(n)]^T$ $\tilde{\mathbf{X}}(n)$ is defined in equation (7.7)	$(L+1)\times 1$	$(M^2+M)/2$
Block C	$\mathbf{C}(n+1) = \mathbf{C}(n) + \mu' e(n)\, \mathbf{S}_{\tilde{Q}}^{-1}\, \tilde{\mathbf{Q}}(n)$	$(L+1)\times 1$	$3(L+1)$
			Total: $2M^2+5M+3$

7.2 Computer Simulation Examples

Example 1.

Consider the system identification problem in figure 7-1 The unknown plant is modeled by a 10-sample memory second-order Volterra series which can be described by equation 7.5:

$$d(n) = h_0 + \sum_{k_1=0}^{9} h_1(k_1)x(n-k_1) + \sum_{k_1=0}^{9}\sum_{k_2=k_1}^{9} h_2(k_1,k_2)x(n-k_1)x(n-k_2) \tag{7.45}$$

where $h_0 = -0.5$, the coefficients of Volterra kernel $h_1(k_1)$ and $h_2(k_1, k_2)$ are shown in Tables 7.2a and 7.2b.

Table 7.2a. Linear coefficients $h_1(k_1)$ for Example 1

$h_1(k_1)$	0	1	2	3	4	5	6	7	8	9
	6.7117e-01	6.2787e-01	4.6008e-01	3.0311e-01	1.8809e-01	1.1231e-01	6.5290e-02	3.7212e-02	2.0889e-02	1.1586e-02

As mentioned in Chapter 3, if the basis functions are given, equation 7.45 may have an equivalent nonlinear Wiener structure which can be expressed by equation 3.91.

$$d(n) = -0.5 + z_0(n) - 0.45\, z_1(n) + z_0^2(n) + 2z_0(n)z_1(n) + 0.2z_1^2(n) \tag{7.46}$$

where $z_0(n) = l_0(n)*x(n)$ and $z_1(n) = l_1(n)*x(n)$. The sign * denotes linear convolution. The $l_0(n)$ and $l_1(n)$ are orthonormal Laguerre filters of length 10 which can be obtained recursively by (Therrien 1993):

Table 7.2b. Quadratic coefficients $h_2(k_1, k_2)$ for Example 1

$h_2(k_1, k_2)$	0	1	2	3	4	5	6	7	8	9
0	1.4625e+00	4.5000e-01	-2.8124e-01	-3.9375e-01	-3.2344e-01	-2.2500e-01	-1.4414e-01	-8.7891e-02	-5.1855e-02	-2.9883e-02
1		-2.2500e-01	-5.6250e-01	-4.5000e-01	-3.0937e-01	-1.9688e-01	-1.1953e-01	-7.0313e-02	-4.0430e-02	-2.2852e-02
2			-2.4610e-01	-3.5156e-01	-2.2852e-01	-1.4062e-01	-8.3496e-02	-4.8340e-02	-2.7466e-02	-1.5381e-02
3				-1.1953e-01	-1.5117e-01	-9.1406e-02	-5.3613e-02	-3.0762e-02	-1.7358e-02	-9.6680e-03
4					-4.7021e-02	-5.6250e-02	-3.2740e-02	-1.8677e-02	-1.0492e-02	-5.8228e-03
5						-1.6700e-02	-1.9336e-02	-1.0986e-02	-6.1523e-03	-3.4058e-03
6							-5.5756e-03	-6.3171e-03	-3.5294e-03	-1.9501e-03
7								-1.7853e-03	-1.9913e-03	-1.0986e-03
8									-5.5447e-04	-6.1111e-04
9										-1.6823e-04

$$l_i(n) = \rho\, l_i(n-1) + \rho l_{i-1}(n-1) - l_{i-1}(n-1) \qquad (7.47)$$

with $l_0(n) = \sqrt{1-\rho^2}\,\rho^n u(n)$ where ρ is a scale factor allowing the flexibility of time-domain scaling. Set $\rho = 0.5$ in this example. The selection of the Laguerre filter is based on its general usefulness in the synthesis of causal operators (King 1977, Silva 1995). Note that a complicated Volterra series can have a much shorter nonlinear Wiener model representation and this representation is unique. However, each term in equation 7.45 still has no complete orthogonality.

To perform the nonlinear LMS algorithm, we need to express equation 7.45 by equation 7.8 as

$$y(n)= c_0 + c_1(0)\widetilde{Q}_0^{(1)}(n) + c_1(1)\widetilde{Q}_1^{(1)}(n) + c_2(0,0)\widetilde{Q}_{00}^{(2)}(n) + c_2(0,1)\widetilde{Q}_{01}^{(2)}(n)$$

$$+ c_2(1,1)\widetilde{Q}_{11}^{(2)}(n) \qquad (7.48)$$

Note that, in equation 7.48, each $\widetilde{Q}$-functional is mutually orthogonal and there are only six coefficients $c_1, c_1(0), c_1(1), c_2(0,0), c_2(0,1), c_2(1,1)$ (that need to be adapted). Therefore, we can run the algorithm as derived and expect good performance.

To perform the nonlinear Wiener LMS algorithm, for the ideal Wiener model (where the orthonormal bases are known), assume the input signal and plant noise are both Gaussian white noise with unit and 10^{-4} variances respectively. Set μ equal to 0.001, and by running 10^4 iterations with 50 independent experiments and averaging over the last 5,000 output data, we have the experimental result of misadjustment MISADJ = 0.0083.

The autocorrelation matrix can be obtained as $\mathbf{R}_{\tilde{Q}\tilde{Q}}$ = diag[1, 1, 1, 2, 1, 2]. Then, we get the theoretical MISADJ = 0.008 that is consistent with the experimental value. If μ is set to be 0.002, with the same condition, we obtain an experimental result of MISADJ = 0.0177 (theoretical value of 0.016). The learning curve of μ = 0.002 is shown in figure 7-3 for the ideal Wiener model. Note that the steady state mean square error has the same power as the plant noise. We also compare this with the learning curves of the Volterra model and the suboptimal Wiener model in figure 7-3. Obviously, both Wiener models have better performance than the Volterra model. This is because the eigenvalue spread is in general large in the Volterra model. We will examine the issue of eigenvalue spread later. To further reduce $\mathbf{R}_{\tilde{Q}\tilde{Q}}$ to an identity matrix, we can find a scale matrix: $\mathbf{S}_{\tilde{Q}}$ = diag $[1, 1, 1, \sqrt{2}, 1, \sqrt{2}]$; then the autocorrelation matrix now becomes an identity matrix: $\mathbf{S}_{\tilde{Q}}^{-2}\mathbf{R}_{\tilde{Q}\tilde{Q}}$ = diag[1, 1, 1, 1, 1, 1]. It is possible to estimate this matrix iteratively. With the same conditions for both μ=0.001 and μ=0.002, the experimental results are MISADJ=0.0062 (theoretical value of 0.006) and 0.0127 (theoretical value of 0.012), respectively. Compared to previous cases, the misadjustment is improved. The summary of the misadjustment is listed in table 7.3.

Table 7.3. MISADJ for different μ values for Example 1

	MISADJ Without $\mathbf{S}_{\tilde{Q}}^{-2}$		MISADJ With $\mathbf{S}_{\tilde{Q}}^{-2}$	
	Experimental value	Theoretical value	Experimental value	Theoretical value
$\mu = 0.001$	0.0083	0.008	0.0062	0.006
$\mu = 0.002$	0.0177	0.016	0.0127	0.012

From above, we see that a two-channel second-order discrete nonlinear Wiener model can represent a 10-sample memory second-order Volterra series. Each channel has 10-sample memory. Theoretically, it allows us to use only 6 coefficients to identify the unknown nonlinear plant which has 66 coefficients.

Practically, however, it is difficult to know the orthonormal bases beforehand. But as we mentioned before, we can select any orthonormal basis to use with the Wiener model to get a suboptimal solution. The most efficient way is to use delay structure, as in figure 7-3. We need 9 delay elements in block A. The Hermite polynomials H_0, H_1, and H_2 are used in block B. For the 10-memory second-order Wiener model, after fully expanding equation 7.8, there are 66 $\tilde{Q}$-polynomials; which means that there are 66 coefficients in block C. Properly select the scale matrix $\mathbf{S}_{\tilde{Q}}$. The learning curve for the suboptimal case is shown in figure 7-2. Compared to the ideal case, it has a slower convergence rate. However, it still has better performance than the Volterra model. The reason is easily seen if we examine the eigenvalue spread of each case.

Referring to table 7.4, the eigenvalue spread for the Volterra model, the ideal Wiener model and the suboptimal Wiener model are 81.343 (compared with theoretical value of 82.488), 1.041 (compared with theoretical value 1) and 1.12 (compared with theoretical value 1), respectively. Comparing with the Volterra model, the eigenvalue spread is reduced about 80 times for either the ideal Wiener model or the suboptimal Wiener model. However, the suboptimal case has a slower convergence rate. We can also compare the autocorrelation matrix of the input vector for each case, as shown in figures 7-4, 7-5, and 7-6. Once again, this shows that both Wiener models have autocorrelation matrices much closer to an identity matrix than the Volterra model.

Table 7.4. Eigenvalue spread characteristics of Example 1

	λ_{max}		λ_{min}		$\lambda_{max} / \lambda_{min}$	
	Experimental value	Theoretical value	Experimental value	Theoretical value	Experimental value	Theoretical value
Wiener Model (ideal case)	1.015	1	0.975	1	1.041	1
Wiener Model (suboptimal case)	1.043	1	0.931	1	1.12	1
Volterra Model	12.119	12.844	0.149	0.156	81.343	82.488

Example 2.

Assume that an arbitrary unknown plant can be described by a 4-sample memory, second-order Volterra series. The input-output relationship is:

$$d(n) = -0.78x(n) - 1.48x(n-1) + 1.39x(n-2) + 0.04x(n-3) + 0.54x^2(n) - 1.62x^2(n-1) + 1.41x^2(n-2) - 0.13x^2(n-3) + 3.72x(n)x(n-1) + 1.86x(n)x(n-2) - 0.76x(n)x(n-3) + 0.76x(n-1)x(n-2) - 0.12x(n-1)x(n-3) - 1.52x(n-2)x(n-3) \quad (7.49)$$

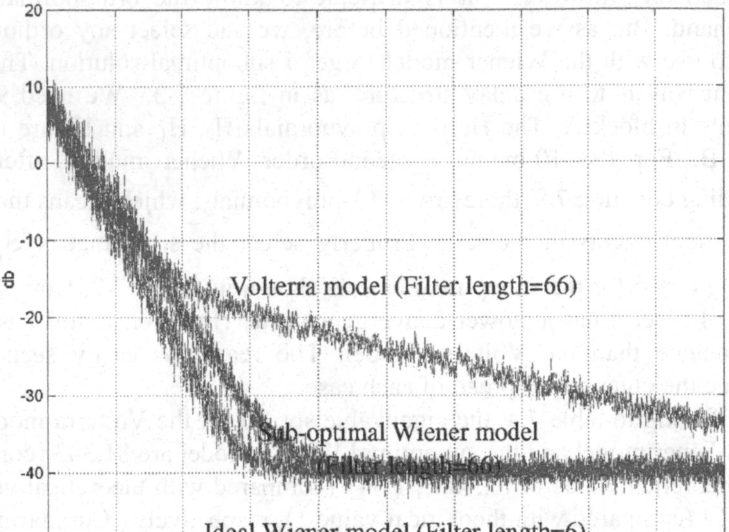

Figure 7-3. Example 1: MSE learning curves with $\mu = 0.002$

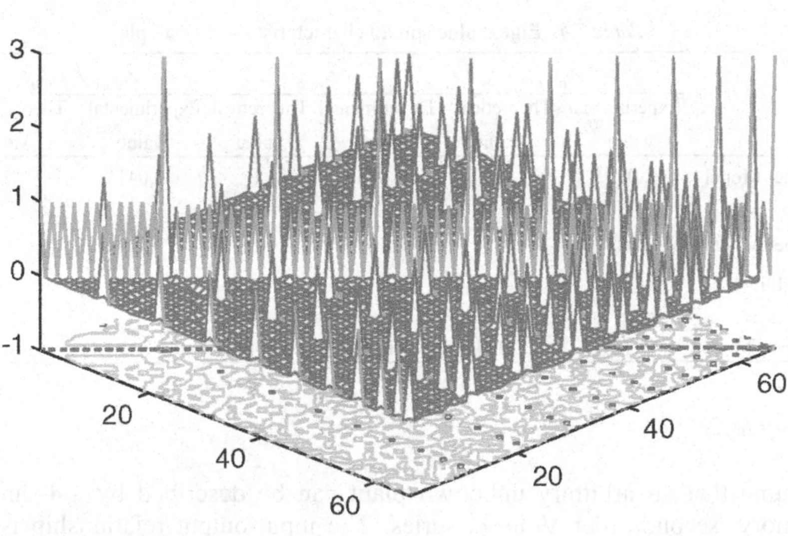

Figure 7-4. Example 1: Autocorrelation matrix (Volterra model)

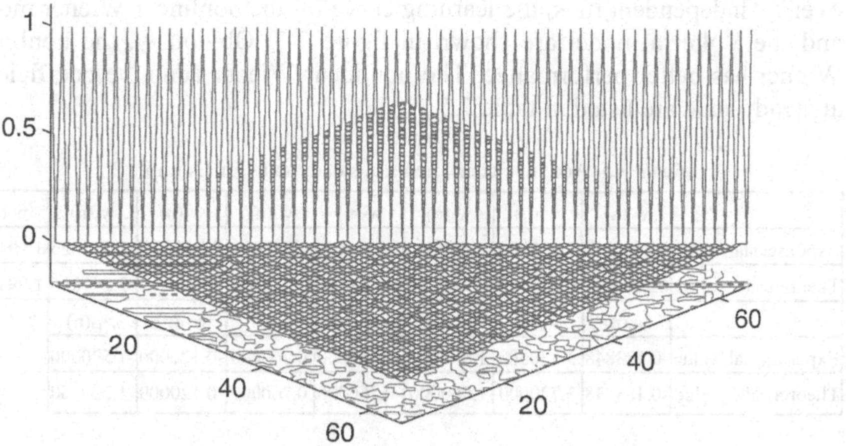

Figure 7-5. Example 1: Autocorrelation matrix (suboptimal Wiener model)

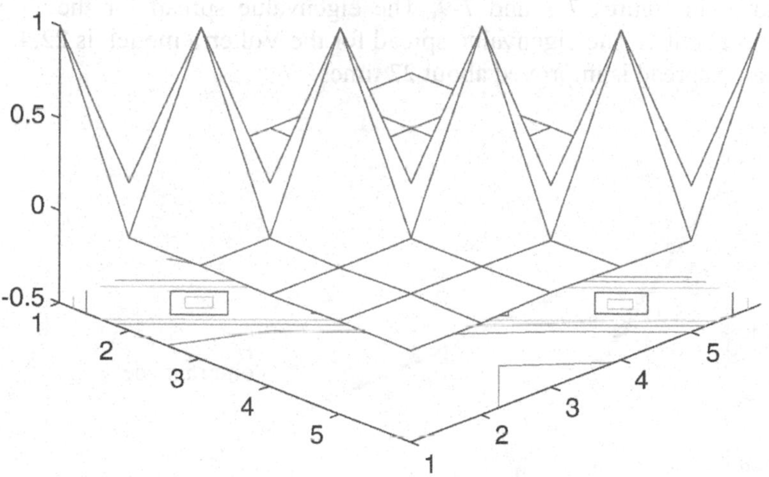

Figure 7-6. Example 1: Autocorrelation matrix (ideal Wiener model)

The input signal x(n) is white Gaussian noise with unit variance. Figure 7-2 structure can be applied in this example again. In block A, there are 3 delay elements used. H_0, H_1, and H_2 are used in block B. For a 4-memory second-order Wiener model, fully expanding equation 7.8, we have a total of 15 $\widetilde{Q}$-polynomials which means that there are 15 coefficients in block C. As in Example 1, properly select the scale matrix $\mathbf{S}_{\tilde{Q}}$ as diag[1,1,1,1,1, $1/\sqrt{2}$, $1/\sqrt{2}$, $1/\sqrt{2}$, $1/\sqrt{2}$,1,1,1,1,1,1], which can let the autocorrelation matrix become an identity matrix. Without adding plant noise, and with ensemble averaging over 50 independent runs, the learning curve of the nonlinear Wiener model and the Volterra model are shown in figure 7-7. Obviously, the nonlinear Wiener has better performance. The nonlinear Wiener adaptive coefficients at steady state are listed in table 7.5.

Table 7.5. Adaptive coefficients in steady state for Example 2

	$w_0(n)$	$w_1(n)$	$w_2(n)$	$w_3(n)$	$w_4(n)$	$w_5(n)$	$w_6(n)$	$w_7(n)$
Experimental value	0.200000	0.780000	-1.480000	1.390000	0.040000	0.763675	-2.291026	1.994041
Theoretical value	0.200000	0.780000	-1.480000	1.390000	0.040000	0.763675	-2.291026	1.994041

	$w_8(n)$	$w_9(n)$	$w_{10}(n)$	$w_{11}(n)$	$w_{12}(n)$	$w_{13}(n)$	$w_{14}(n)$
Experimental value	-0.183848	3.720000	1.860000	-0.760000	0.760000	-0.120000	-1.520000
Theoretical value	-0.183848	3.720000	1.860000	-0.760000	0.760000	-0.120000	-1.520000

We note that all the values in table 7.5 match the theoretical values, which are derived in chapter 3. To examine the eigenvalue spread characteristics, we need to evaluate the autocorrelation matrix. The experimental auto-correlation matrices for the nonlinear Wiener model and the Volterra model are shown in figures 7-8 and 7-9. The eigenvalue spread for the Wiener model is about 1. The eigenvalue spread for the Volterra model is 22.4. The eigenvalue spread is improved about 22 times.

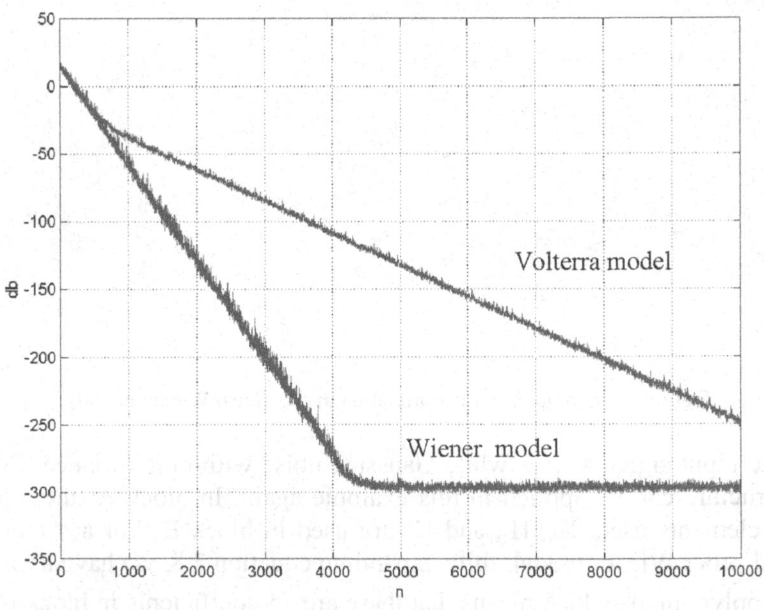

Figure 7-7. Example 2: MSE Learning Curves

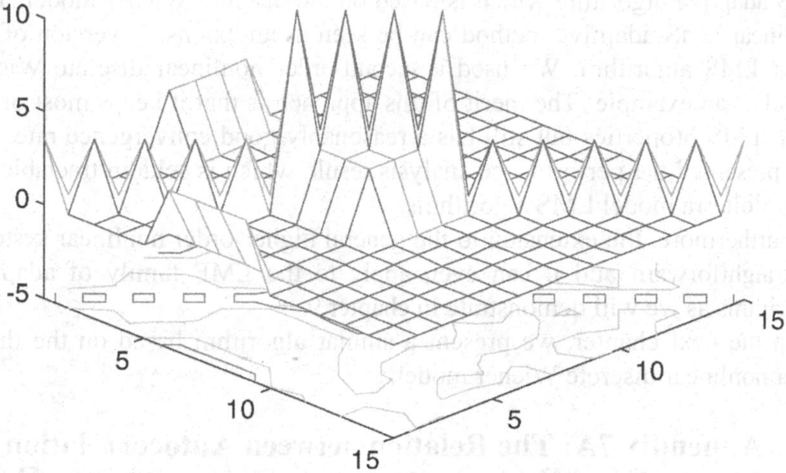

Figure 7-8. Example 2: Autocorrelation matrix (Volterra model)

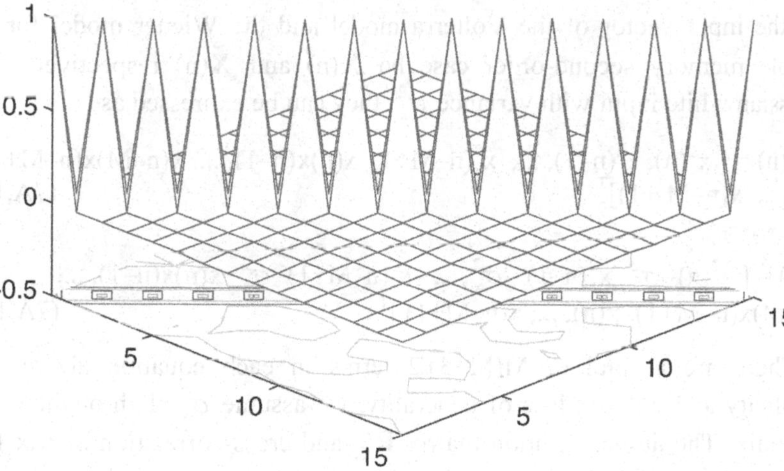

Figure 7-9. Example 2: Autocorrelation matrix (Wiener model)

7.3 Summary

In this chapter, we extended the developments of chapter 6 to a nonlinear LMS adaptive algorithm which is based on the discrete Wiener model. This nonlinear LMS adaptive method can be seen as an extension version of the linear LMS algorithm. We used a second-order nonlinear discrete Wiener model as an example. The merit of this approach is that it keeps most of the linear LMS properties but still has a reasonably good convergence rate. We also presented the performance analysis result, which is seldom tractable for most Volterra model LMS algorithms.

Furthermore, the extension to the general higher-order nonlinear systems is straightforward and it can even apply to the LMF family of adaptive algorithms as we will demonstrate in chapter 9.

In the next chapter, we present a similar algorithm based on the third-order nonlinear discrete Wiener model.

7.4 Appendix 7A: The Relation between Autocorrelation Matrix $\mathbf{R}_{\mathbf{xx}}$, $\mathbf{R}_{\tilde{\mathbf{x}}\tilde{\mathbf{x}}}$ and Cross-Correlation Matrix $\mathbf{R}_{\tilde{\mathbf{x}}\mathbf{x}}$

7.4.1 Second-Order Case

Let the input vector of the Volterra model and the Wiener model for M-sample memory second-order case be $\mathbf{X}(n)$ and $\tilde{\mathbf{X}}(n)$ respectively. For Gaussian white input with variance σ_x^2, they can be expressed as

$$\mathbf{X}(n) = [\, x^2(n), x^2(n-1), \ldots, x^2(n-M+1), x(n)x(n-1), \ldots, x(n-M)x(n-M+1), x(n), \ldots, x(n-M+1)]^T \tag{7A.1.1}$$

$$\tilde{\mathbf{X}}(n) = [\, x^2(n)-\sigma_x^2, x^2(n-1)-\sigma_x^2, \ldots, x^2(n-M+1)-\sigma_x^2, x(n)x(n-1), \ldots, x(n-M)x(n-M+1), x(n), \ldots, x(n-M+1)\,]^T \tag{7A.1.2}$$

There are a total of M(M+3)/2 terms in each equation above. For simplicity and without loss of generality, we assume $\sigma_x^2=1$ throughout this appendix. The autocorrelation matrix $\mathbf{R}_{\tilde{\mathbf{x}}\tilde{\mathbf{x}}}$ and cross-correlation matrix $\mathbf{R}_{\tilde{\mathbf{x}}\mathbf{x}}$ are defined as

$$\mathbf{R}_{\tilde{\mathbf{x}}\tilde{\mathbf{x}}} = E\{\tilde{\mathbf{X}}(n)\tilde{\mathbf{X}}^T(n)\} \tag{7A.1.3}$$

$$\mathbf{R}_{\tilde{\mathbf{x}}\mathbf{x}} = E\{\tilde{\mathbf{X}}(n)\mathbf{X}^T(n)\} \tag{7A.1.4}$$

$\mathbf{R}_{\tilde{X}\tilde{X}}$ is a diagonal matrix, because we showed in chapter 3 that the Q-polynomial set satisfies the self-orthogonality property. In fact, the expression of $\mathbf{R}_{\tilde{X}\tilde{X}}$ can be shown to be

$$\mathbf{R}_{\tilde{X}\tilde{X}} = \mathrm{diag}[\underbrace{2,\ 2,\ \ldots,\ 2,}_{M(M-1)/2\text{'s}}\ \underbrace{1,\ 1,\ \ldots,\ 1}_{2M\text{'s}}] \tag{7A.1.5}$$

To obtain equation 7A.1.5, we use two facts. The first fact is that the mean value is zero for odd number's of Gaussian white random variables which means that

$$E\{x(n-k_0)x(n-k_1)x(n-k_2)\} = 0 \tag{7A.1.6a}$$

The second fact is based on the well known fourth-order joint moment of Gaussian random variables (Papoulis 1991) which is

$$\begin{aligned} &E\{x(n-k_0)x(n-k_1)x(n-k_2)x(n-k_3)\} \\ &\quad = E\{x(n-k_0)x(n-k_1)\}E\{x(n-k_2)x(n-k_3)\} \\ &\quad + E\{x(n-k_0)x(n-k_2)\}E\{x(n-k_1)x(n-k_3)\} \\ &\quad + E\{x(n-k_0)x(n-k_3)\}\ E\{x(n-k_2)x(n-k_3)\} \end{aligned} \tag{7A.1.6b}$$

From equation 7A.1.6a and 7A.1.6b, we can verify that equation 7A.1.5 is true. For instance,

$$\begin{aligned} &E\{(x^2(n-k_0)-1)(x^2(n-k_0)-1)\} \\ &\qquad = E\{x^4(n-k_0)\} - 2E\{x^2(n-k_0)\} + 1 = 2 \end{aligned} \tag{7A.1.7a}$$

$$E\{(x^2(n-k_0)-1)x^2(n-k_1)\} = 0, \quad \text{for } k_0 \neq k_1 \tag{7A.1.7b}$$

For cross-correlation matrix $\mathbf{R}_{\tilde{X}X}$, this is a diagonal matrix too, which is

$$\mathbf{R}_{\tilde{X}X} = \mathrm{diag}[\underbrace{2,\ 2,\ \ldots,\ 2,}_{M(M-1)/2\text{'s}}\ \underbrace{1,\ 1,\ \ldots,\ 1}_{2M\text{'s}}\ ] \tag{7A.1.8}$$

Equation 7A.1.8 can be quickly verified by the facts of equation 7A.1.6a, 7A.1.6b, and

$$E\{(x^2(n-k_0)-1)x^2(n-k_0)\} = E\{x^4(n-k_0)\} - E\{x^2(n-k_0)\} = 3 - 1 = 2 \tag{7A.1.9}$$

From equation 7A.1.5 and equation 7A.1.8, we know immediately that the autocorrelation matrix $\mathbf{R}_{\tilde{X}\tilde{X}}$ and the cross-correlation matrix $\mathbf{R}_{\tilde{X}X}$ are identical; i.e.,

$$\mathbf{R}_{\tilde{X}X} = \mathbf{R}_{\tilde{X}\tilde{X}} \qquad (7A.1.10)$$

To explore the relationship between the two autocorrelation matrices, we need to define $\mathbf{m}_x = E\{\mathbf{X}(n)\}$, which implies that $\mathbf{m}_x = [E\{x^2(n)\}, E\{x^2(n-1)\}, \ldots, E\{x^2(n-M+1)\}, E\{x(n)x(n-1)\}, \ldots, E\{x(n-M)x(n-M+1)\}, E\{x(n)\}, \ldots, E\{x(n-M+1)\}]^T$

$$= [1, 1, \ldots, 1, 0, \ldots, 0, 0, \ldots, 0]^T \qquad (7A.1.11)$$

From equation 7A.1.11, it is not difficult to show that

$$\mathbf{m}_x\mathbf{m}_x^T = \begin{bmatrix} \mathbf{1} & \mathbf{0} \\ \mathbf{0} & \mathbf{1} \end{bmatrix} \qquad (7A.1.12)$$

where **1** is a square matrix containing all 1's components and **0** are the matrices containing all 0's components. Based on the facts of equation 7A.1.6a, equation 7A.1.6, and equation 7A.1.9, $\mathbf{R}_{XX}$ can be expressed as

$$\mathbf{R}_{XX} = \begin{bmatrix} 3 & 1 & \cdots & 1 & & & \\ 1 & 3 & \cdots & 1 & \mathbf{0} & & \\ & & \ddots & & & & \\ 1 & 1 & \cdots & 3 & & & \\ & & & & 1 & & \\ & \mathbf{0} & & & & 1 & \\ & & & & & & 1 \end{bmatrix} \qquad (7A.1.13)$$

Note that the dimension of the up-right corner matrix is the same as **1** of equation 7A.1.12. Combine equation 7A.1.5, equation 7A.1.12, and equation 7A.1.13, and we can express $\mathbf{R}_{XX}$ in terms of $\mathbf{R}_{\tilde{X}X}$ and $\mathbf{m}_x$, which is

$$\mathbf{R}_{XX} = \mathbf{R}_{\tilde{X}X} + \mathbf{m}_x\mathbf{m}_x^T \qquad (7A.1.14)$$

For the second-order case, we can see that the relation between $\mathbf{R}_{XX}$, $\mathbf{R}_{\tilde{X}X}$ and $\mathbf{R}_{\tilde{X}\tilde{X}}$ is clearly described by equation 7A.1.10 and equation 7A.1.14.

7.5 Appendix 7B: General-Order Moments of Joint Gaussian Random Variables

Assume that x is a zero mean, unit variance Gaussian random variable. The result that we want to develop is the general form of expectation of jointly

Gaussian random variables which is the $(n_1+n_2+...+n_k)$th moment value of $m_{n_1+n_2+...+n_k}$

$$m_{n_1+n_2+...+n_k} = E\{x_1^{n_1} x_2^{n_2} \ldots x_k^{n_k}\}, \quad \text{where} \quad n_1+n_2+...+n_k = \text{even} \tag{7B.1}$$

For $n_1+n_2+...+n_k$ = odd, equation 7B.1 is equal to zero. To find equation 7B.1, we can apply the property of characteristic functions which is defined as (Papoulis 1991):

$$\phi_X(\mathbf{w}) = E\{e^{j\mathbf{w}^T\mathbf{x}}\} \tag{7B.2}$$

where $\mathbf{w} = [\ w_1, w_2, ..., w_k\ ]^T$ and $\mathbf{x} = [x_1, x_2, ...\ x_k]^T$. Equation 7B.2 is a function of real numbers, $-\infty \leq w_i \leq \infty$. Before showing the general form of (7B.1), we first derive the special case of (n_1+n_2) th moment value for two random variables which is

$$m_{n_1+n_2} = E\{x_1^{n_1} x_2^{n_2}\}, \qquad n_1+n_2 = \text{even} \tag{7B.3}$$

According to equation 7B.2, the characteristic function for two random variables is

$$\phi_X(\mathbf{w}) = E\{e^{j\mathbf{w}^T\mathbf{x}}\} = E\{e^{j(n_1 w_1 x_1 + n_2 w_2 x_2)}\} \tag{7B.4}$$

where $\mathbf{w} = [w_1, ..., w_1, w_2, ..., w_2]^T$ and $\mathbf{x} = [x_1, ..., x_1, x_2, ...\ , x_2]^T$. The (n_1+n_2)th moment can be obtained as (Peebles 1987):

$$m_{n_1+n_2} = (-j)^{n_1+n_2} \left. \frac{d^{n_1+n_2} \phi_X(\mathbf{w})}{d^{n_1}w_1 d^{n_2}w_2} \right|_{w_1=w_2=0}$$

$$=(-j)^{n_1+n_2} \frac{d^{n_1+n_2}}{d^{n_1}w_1 d^{n_2}w_2} \left(E\{1+j(n_1w_1x_1+n_2w_2x_2)+...+ \frac{[j(n_1w_1x_1+n_2w_2x_2)]^{n_1+n_2}}{(n_1+n_2)!}+...\}\right)\Big|_{w_1=w_2=0}$$

$$= \frac{\binom{n_1+n_2}{n_1} n_1^{n_1} n_2^{n_2} n_1! n_2!}{(n_1+n_2)!} E\{x_1^{n_1} x_2^{n_2}\} = n_1^{n_1} n_2^{n_2} E\{x_1^{n_1} x_2^{n_2}\} \tag{7B.5}$$

For the zero mean, unit variance Gaussian random variable, we know (from Peebles 1987) that

$$\phi_X(\mathbf{w}) = e^{j\mathbf{w}^T E(\mathbf{x}\mathbf{x}^T)\mathbf{w}} = \exp\Big(\tfrac{-1}{2}[w_1 \ldots w_1, w_2 \ldots w_2]\begin{bmatrix} C_{11} & \ldots & C_{11} & C_{12} & \ldots & C_{12} \\ & \ldots & & & \ldots & \\ C_{11} & \ldots & C_{11} & C_{12} & \ldots & C_{12} \\ C_{21} & \ldots & C_{21} & C_{22} & \ldots & C_{22} \\ & \ldots & & & \ldots & \\ C_{21} & \ldots & C_{21} & C_{22} & \ldots & C_{22} \end{bmatrix}\begin{bmatrix} w_1 \\ \vdots \\ w_1 \\ w_2 \\ \vdots \\ w_2 \end{bmatrix}\Big) \quad \begin{matrix} \}\, n_1 \text{ terms} \\ \}\, n_2 \text{ terms} \end{matrix}$$

$$= 1 + \left(\tfrac{-1}{2}\right)\left(n_1^2 C_{11} w_1^2 + n_2^2 C_{22} w_2^2 + 2 n_1 n_2 C_{12} w_1 w_2\right)/1! + \ldots$$

$$+ \left(\left(\tfrac{-1}{2}\right)\left(n_1^2 C_{11} w_1^2 + n_2^2 C_{22} w_2^2 + 2 n_1 n_2 C_{12} w_1 w_2\right)\right)^{(n_1+n_2)/2} / \left(\tfrac{n_1+n_2}{2}\right)! + \ldots \quad (7B.6)$$

where $C_{ij} = \text{cov}\{x_i x_j\}$. Then, by using binomial and multinomial theorems (Broshtein 1985), we can find:

$m_{n_1+n_2}$ as

$$m_{n_1+n_2} = (-j)^{n_1+n_2} \left.\frac{d^{n_1+n_2}(Eq.7B.6)}{d^{n_1} w_1 d^{n_2} w_2}\right|_{w_1=w_2=0}$$

$$= (-j)^{n_1+n_2} \left(\tfrac{-1}{2}\right)^{(n_1+n_2)/2} \frac{1}{\left(\frac{n_1+n_2}{2}\right)!}$$

$$\sum_{m_1+m_2+m_3=\frac{n_1+n_2}{2}} \begin{pmatrix} & \frac{n_1+n_2}{2} & \\ m_1 & m_2 & m_3 \end{pmatrix} n_1^{2m_1} n_2^{2m_2} (2 n_1 n_2)^{m_3} n_1! n_2! \, C_{11}^{m_1} C_{22}^{m_2} C_{12}^{m_3}$$

(7B.7)

where $\begin{pmatrix} & \frac{n_1+n_2}{2} & \\ m_1 & m_2 & m_3 \end{pmatrix} = \frac{\left(\frac{n_1+n_2}{2}\right)!}{m_1! m_2! m_3!}$. Comparing 7B.5 and 7B.7, we can obtain the general form of $E\{x_1^{n_1} x_2^{n_2}\}$ as:

$$E\{x_1^{n_1} x_2^{n_2}\} = (-j)^{n_1+n_2} \frac{\left(\frac{-1}{2}\right)^{(n_1+n_2)/2} \frac{1}{\left(\frac{n_1+n_2}{2}\right)!}}{n_1^{n_1} n_2^{n_2}} n_1! n_2!$$

$$\sum_{m_1+m_2+m_3=\frac{n_1+n_2}{2}} \begin{pmatrix} & \frac{n_1+n_2}{2} & \\ m_1 & m_2 & m_3 \end{pmatrix} n_1^{2m_1} n_2^{2m_2} (2n_1n_2)^{m_3} n_1!n_2!\ C_{11}^{m_1} C_{22}^{m_2} C_{12}^{m_3} \tag{7B.8}$$

Now we can derive the general form of $(n_1+n_2+...+n_k)$th moment of jointly Gaussian random variables. As in equation 7B.4, we can have

$$m_{n_1+n_2+...+n_k} = (-j)^{n_1+n_2+...+n_k} \left. \frac{d^{n_1+n_2+...+n_k} \phi_X(\mathbf{w})}{d^{n_1}w_1 d^{n_2}w_2 \ldots d^{n_k}w_n} \right|_{w_1=w_2=...w_k=0}$$

$$= (-j)^{n_1+n_2+...+n_k} \frac{d^{n_1+n_2+...+n_k}}{d^{n_1}w_1 d^{n_2}w_2 \ldots d^{n_k}w_k} (E\{1+...$$

$$\left. + \frac{[j(n_1w_1x_1 + n_2w_2x_2 + ... + n_kw_kx_k)]^{n_1+n_2+...+n_k}}{(n_1+n_2+...+n_k)!} + ... \}) \right|_{w_1=w_2=...w_k=0}$$

$$= \frac{\begin{pmatrix} \frac{n_1+n_2+...+n_k}{2} \\ n_1 \quad \ldots \quad n_k \end{pmatrix} n_1^{n_1} \ldots n_k^{n_k} n_1!n_2! \ldots n_k!}{(n_1+n_2+...+n_k)!} E\{x_1^{n_1} x_2^{n_2} \ldots x_k^{n_k}\}$$

$$= n_1^{n_1} \ldots n_k^{n_k} E\{x_1^{n_1} x_2^{n_2} \ldots x_k^{n_k}\} \tag{7B.9}$$

For zero mean, unit variance Gaussian random variables, as in equation 7B.6, we know that:

$$\phi_X(\mathbf{w}) = e^{j\mathbf{w}^T E(\mathbf{xx}^T)\mathbf{w}}$$

$$= \exp\left(\frac{-1}{2}[w_1 \ldots w_1 \ldots w_k \ldots w_k] \begin{bmatrix} C_{11} & \ldots & C_{11} & \ldots & C_{1k} & \ldots & C_{1k} \\ & \ldots & & & & \ldots & \\ C_{11} & \ldots & C_{11} & \ldots & C_{1k} & \ldots & C_{1k} \\ & & & \ldots & & & \\ C_{k1} & \ldots & C_{k1} & \ldots & C_{kk} & \ldots & C_{kk} \\ & \ldots & & \ldots & & \ldots & \\ C_{k1} & \ldots & C_{k1} & \ldots & C_{kk} & \ldots & C_{kk} \end{bmatrix} \begin{bmatrix} w_1 \\ \vdots \\ w_1 \\ \vdots \\ w_k \\ \vdots \\ w_k \end{bmatrix}\right)$$

$$=\exp[\tfrac{-1}{2}(n_1^2 C_{11}w_1^2 + ... + n_k^2 C_{kk}w_k^2 + 2n_1n_2C_{12}w_1w_2 + ... + 2n_{k-1}n_kC_{k-1,k}w_{k-1}w_k)]$$

$= 1+\ldots$

$+\left(\left(\frac{-1}{2}\right)\left(n_1^{\,2}C_{11}w_1^2+\ldots+n_k^{\,2}C_{kk}w_k^2+2n_1n_2C_{12}w_1w_2+\ldots+2n_{k-1}n_kC_{k-1,k}w_{k-1}w_k\right)\right)^{\frac{n_1+n_2+\ldots+n_k}{2}} /$

$\left(\frac{n_1+n_2+\ldots n_k}{2}\right)! +\ldots$ (7B.10)

where C_{ij} = $\mathrm{cov}\{x_i x_j\}$. As in equation 7B.7, we can also find the $(n_1+n_2+\ldots+n_k)$th moment:

$$m_{n_1+n_2+\ldots+n_k} = (-j)^{n_1+n_2+\ldots+n_k} \left.\frac{d^{n_1+n_2+\ldots+n_k}(Eq.7B.10)}{d^{n_1}w_1 d^{n_2}w_2\ldots d^{n_k}w_k}\right|_{w_1=w_2=\ldots=w_k=0}$$

$$= (-j)^{n_1+n_2+\ldots+n_k}\left(\frac{-1}{2}\right)^{\frac{n_1+n_2+\ldots+n_k}{2}} \frac{1}{\left(\frac{n_1+n_2+\ldots+n_k}{2}\right)!}$$

$$\sum_{m_1+\ldots+m_m=\frac{n_1+\ldots+n_k}{2}} \begin{pmatrix} & \frac{n_1+\ldots+n_k}{2} & \\ m_1 & \ldots & m_m \end{pmatrix}$$

$$n_1^{\,2m_1}\ldots n_n^{\,2m_n}(2n_1n_2)^{m_{n+1}}\ldots(2n_kn_{k-1})^{m_m}\, n_1!n_2!\ldots n_k!$$

$$C_{11}^{m_1}\ldots C_{kk}^{m_n}C_{12}^{m_{n+1}}\ldots C_{k-1,k}^{m_m} \qquad (7B.11)$$

Compare equation 7B.9 and equation 7B.11, we find that

$$E\{x_1^{n_1}x_2^{n_2}\ldots x_k^{n_k}\} = (-j)^{n_1+n_2+\ldots+n_k}\frac{\left(\frac{-1}{2}\right)^{\frac{n_1+n_2+\ldots+n_k}{2}}\frac{1}{\left(\frac{n_1+n_2+\ldots+n_k}{2}\right)!}}{n_1^{n_1}\ldots n_k^{n_k}} n_1!\, n_2!\ldots n_k! \qquad (7B.12)$$

$$\sum_{m_1+\ldots+m_m=\frac{n_1+\ldots+n_k}{2}} \begin{pmatrix} & \frac{n_1+\ldots+n_k}{2} & \\ m_1 & \ldots & m_m \end{pmatrix} n_1^{2m_1}\ldots n_k^{2m_n}(2n_1n_2)^{m_{n+1}}\ldots(2n_kn_{k-1})^{m_m}$$

$$C_{11}^{m_1}\ldots C_{kk}^{m_n}C_{12}^{m_{n+1}}\ldots C_{k-1,k}^{m_m}$$

Equation 7B.12 is the general form for $(n_1+n_2+\ldots+n_k)$th moment value of 7B.1.

Example: For an application, let us find the value of $E\{x_1x_2^2x_3^3\}$.
From 7B.9, we have:

$$\phi_X(\mathbf{w}) = E\{e^{j(w_1x_1+2w_2x_2+3w_nx_n)}\} = 2^2 3^3 E\{x_1x_2^2x_3^3\} \qquad (7B.13)$$

From 7B.10, we obtain:

$$\phi_X(\mathbf{w}) = e^{j\mathbf{w}^T E(\mathbf{xx}^T)w} = \exp(\tfrac{-1}{2}[w_1\, w_2\, w_2\, w_3\, w_3\, w_3]$$

$$\begin{bmatrix} C_{11} & C_{12} & C_{12} & C_{13} & C_{13} & C_{13} \\ C_{21} & C_{22} & C_{22} & C_{23} & C_{23} & C_{23} \\ C_{21} & C_{22} & C_{22} & C_{23} & C_{23} & C_{23} \\ C_{31} & C_{32} & C_{32} & C_{33} & C_{33} & C_{33} \\ C_{31} & C_{32} & C_{32} & C_{33} & C_{33} & C_{33} \\ C_{31} & C_{32} & C_{32} & C_{33} & C_{33} & C_{33} \end{bmatrix} \begin{bmatrix} w_1 \\ w_2 \\ w_2 \\ w_3 \\ w_3 \\ w_3 \end{bmatrix})$$

$$=\exp[\tfrac{-1}{2}(C_{11}w_1^2+4C_{22}w_2^2+9C_{33}w_3^2+2\cdot 1\cdot 2C_{12}w_1w_2+2\cdot 1\cdot 3C_{13}w_1w_3+2\cdot 2\cdot 3C_{23}w_2w_3)]$$

$$= 1+\ldots \tag{7B.14}$$

$$+\left(\left(\tfrac{-1}{2}\right)\left(C_{11}w_1^2+4C_{22}w_2^2+9C_{33}w_3^2+4C_{12}w_1w_2+6C_{13}w_1w_3+12C_{23}w_2w_3\right)\right)^3/3!+\ldots$$

Using 7B.11, we can find the 6th-moment as:

$$m_6 = (-j)^6 \left.\frac{d^6(C.14)}{dw_1 d^2 w_2 d^3 w_3}\right|_{w_1=w_2=w_3=0} \tag{7B.15}$$

$$=(-j)^6(\tfrac{-1}{2})^3\tfrac{1}{3!}\ \Big(\sum_{m_1+\ldots+m_6=3} \binom{3}{m_1\ \ldots\ m_6} 1^{2m_1}2^{2m_2}3^{2m_3}4^{m_4}6^{m_5}\ 12^{m_6}1!2!3!C_{11}^{m_1}C_{22}^{m_2} C_{33}^{m_3}C_{12}^{m_4}C_{13}^{m_5}C_{23}^{m_6}\Big)$$

Compare 7B.13 with 7B.15. We obtain

$E\{x_1x_2^2x_3^3\} =$

$$j^6\frac{(\frac{-1}{2})^3\frac{1}{3!}}{1^1 2^2 3^3}\Big(\binom{3}{001101}\cdot 4\cdot 12\cdot 9\cdot 1!\cdot 2!\cdot 3!\ C_{12}C_{23}C_{33}+\binom{3}{011010}\cdot 6\cdot 4\cdot 9\cdot 1!\cdot 2!\cdot 3!\ C_{13}C_{22}C_{33}$$

$$+\begin{pmatrix} 3 \\ 000012 \end{pmatrix} \cdot 6 \cdot 12 \cdot 12 \cdot 1! \cdot 2! \cdot 3! C_{13} C_{23}^2) = 6C_{12} C_{23} C_{33}$$

$$+ 3C_{13} C_{22} C_{33} + 6C_{13} C_{23}^2 \tag{7B.16}$$

For zero mean, unit variance white Gaussian random variable, if $x_1 = x_2 = x_3$, then

$$E\{ x_1^6 \} = 15 \tag{7B.17}$$

The answer to 7B.17 can be verified by (Schetzen 1980):

$$E\{x^{2n}\}_{n=3} = \frac{(2 \cdot 3)!}{3! \cdot 2^3} = 15 \tag{7B.18}$$

This procedure can be applied directly to any general-order condition. *Example:*

$E\{x_1^3 x_2 x_3^2 x_4^4\}$

$$= (-j)^{10} \frac{(\frac{-1}{2})^5 \frac{1}{5!}}{3^3 1^1 2^2 4^4} 3!1!2!4!$$

$$(\begin{pmatrix} 5 \\ 1\,0\,1\,2\,1\,0\,0\,0\,0\,0 \end{pmatrix} \cdot 9 \cdot 4 \cdot 16 \cdot 16 \cdot 6\, C_{11}C_{33}C_{44}C_{44}C_{12}$$

$$+\begin{pmatrix} 5 \\ 1\,0\,0\,2\,0\,1\,0\,1\,0\,0 \end{pmatrix} \cdot 9 \cdot 16 \cdot 16 \cdot 12 \cdot 4\, C_{11}C_{44}C_{44}C_{13}C_{23}$$

$$+\begin{pmatrix} 5 \\ 1\,0\,1\,1\,0\,0\,1\,0\,1\,0 \end{pmatrix} \cdot 9 \cdot 4 \cdot 16 \cdot 24 \cdot 8\, C_{11}C_{33}C_{44}C_{14}C_{24}$$

$$+\begin{pmatrix} 5 \\ 1\,0\,0\,1\,0\,1\,0\,0\,1\,1 \end{pmatrix} \cdot 9 \cdot 16 \cdot 12 \cdot 8 \cdot 16\, C_{11}C_{44}C_{13}C_{24}C_{34}$$

$$+\begin{pmatrix} 5 \\ 1\,0\,0\,1\,0\,0\,1\,1\,0\,1 \end{pmatrix} \cdot 9 \cdot 16 \cdot 24 \cdot 4 \cdot 16\, C_{11}C_{44}C_{14}C_{23}C_{34}$$

$$+\begin{pmatrix} 5 \\ 0\,0\,0\,2\,1\,2\,0\,0\,0\,0 \end{pmatrix} \cdot 16 \cdot 16 \cdot 12 \cdot 12 \cdot 6\, C_{44}C_{44}C_{13}C_{13}C_{12}$$

$$+\begin{pmatrix} 5 \\ 0\,0\,1\,1\,1\,0\,2\,0\,0\,0 \end{pmatrix} \cdot 4 \cdot 16 \cdot 24 \cdot 24 \cdot 6\, C_{33}C_{44}C_{14}C_{14}C_{12}$$

$$+\begin{pmatrix} 5 \\ 1\,0\,0\,1\,1\,0\,0\,0\,0\,2 \end{pmatrix} \cdot 9 \cdot 16 \cdot 16 \cdot 16 \cdot 6\, C_{11}C_{44}C_{34}C_{34}C_{12}$$

$$+\begin{pmatrix} 5 \\ 0\,0\,0\,1\,1\,1\,1\,0\,0\,1 \end{pmatrix} \cdot 16 \cdot 12 \cdot 16 \cdot 24 \cdot 6\, C_{44}C_{13}C_{34}C_{14}C_{12}$$

$$+\binom{5}{0\,0\,1\,0\,0\,0\,3\,0\,1\,0}\cdot 4\cdot 24\cdot 24\cdot 24\cdot 8\, C_{33}C_{14}C_{14}C_{14}C_{24}$$

$$+\binom{5}{0\,0\,0\,1\,0\,2\,1\,0\,1\,0}\cdot 16\cdot 12\cdot 12\cdot 24\cdot 8\, C_{44}C_{13}C_{13}C_{14}C_{24}$$

$$+\binom{5}{0\,0\,0\,1\,0\,1\,2\,1\,0\,0}\cdot 16\cdot 24\cdot 24\cdot 12\cdot 4\, C_{44}C_{14}C_{14}C_{13}C_{23}$$

$$+\binom{5}{1\,0\,0\,0\,0\,0\,1\,0\,1\,2}\cdot 9\cdot 16\cdot 16\cdot 24\cdot 8\, C_{11}C_{34}C_{34}C_{14}C_{24}$$

$$+\binom{5}{0\,0\,0\,0\,1\,0\,2\,0\,0\,2}\cdot 24\cdot 24\cdot 16\cdot 16\cdot 6\, C_{14}C_{14}C_{34}C_{34}C_{12}$$

$$+\binom{5}{0\,0\,0\,0\,0\,1\,2\,0\,1\,1}\cdot 12\cdot 16\cdot 24\cdot 24\cdot 8\, C_{13}C_{34}C_{14}C_{14}C_{24}$$

$$+\binom{5}{0\,0\,0\,0\,0\,0\,3\,1\,0\,1}\cdot 24\cdot 24\cdot 4\cdot 16\cdot 24\, C_{14}C_{14}C_{23}C_{34}C_{14}) \qquad (7B.19)$$

Therefore,

$$E\{x_1^3x_2x_3^2\ x_4^4\} = 9C_{11}C_{33}C_{44}C_{44}C_{12} + 8C_{11}C_{44}C_{44}C_{13}C_{23} + 36C_{11}C_{33}C_{44}C_{14}C_{24} + 72C_{11}C_{44}C_{13}C_{24}C_{34} + 72\,C_{11}C_{44}C_{14}C_{23}C_{34} + 18C_{44}C_{44}C_{13}C_{13}C_{12} + 36C_{33}C_{44}C_{14}C_{14}C_{12} + 36C_{11}C_{44}C_{34}C_{34}C_{12} + 144C_{44}C_{13}C_{34}C_{14}C_{12} + 24C_{33}C_{14}C_{14}C_{14}C_{24} + 72C_{44}C_{13}C_{13}C_{14}C_{24} + 72C_{44}C_{14}C_{14}C_{13}C_{23} + 72C_{11}C_{34}C_{34}C_{14}C_{24} + 72C_{14}C_{14}C_{34}C_{34}C_{12} + 144C_{13}C_{34}C_{14}C_{14}C_{24} + 48\,C_{14}C_{14}C_{23}C_{34}C_{14} \qquad (7B.20)$$

For zero mean, unit variance white Gaussian random variable, if $x_1 = x_2 = x_3 = x_4 = x$, then:

$$E\{x^{10}\} = 9+18+36+72+72+18+36+36+144+24+72+72+72+72+144+48 = 945 \qquad (7B.21)$$

The answer to 7B.21 can be verified by:

$$E\{x^{2n}\}_{n=5} = \frac{10!}{5!2^5} = 945 \qquad (7B.22)$$

Chapter 8

NONLINEAR ADAPTIVE SYSTEM IDENTIFICATION BASED ON WIENER MODELS (PART 2)

Third-order least-mean-square (LMS)-based approach

Introduction

In this chapter, we study using third-order nonlinear Wiener models for for adaptive system identification. As in chapter 7, we introduce a self-orthogonalizing method that is based on a delay-line structure of the nonlinear discrete-time Wiener model.

8.1 Third-Order System

We wish to develop an adaptive algorithm based on the third-order nonlinear Wiener model system shown in figure 8-1:

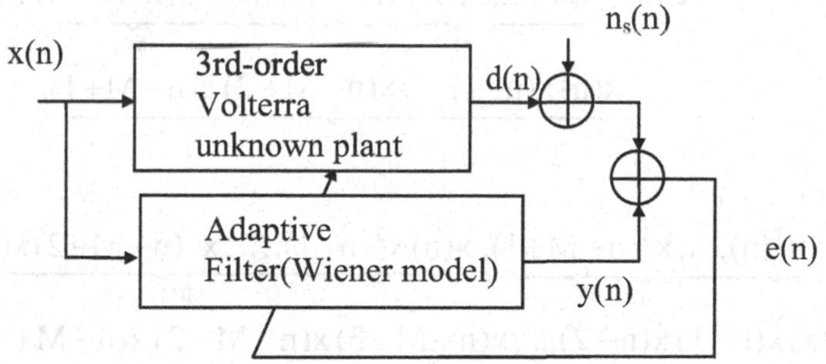

Figure 8-1. Third-order nonlinear system identification model

The unknown plant is a third-order Volterra system (Li 1998, Li 1996, Im 1996). To be consistent with the previous chapter, we apply the delay-line version of the third-order discrete nonlinear Wiener model in the adaptive plant. The block diagram is shown in figure 8-2.

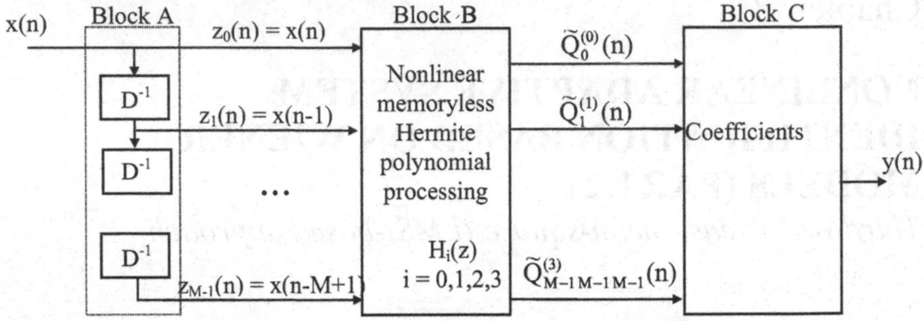

Figure 8-2. Delay line structure of a third-order nonlinear discrete Wiener model

As the order increases, the complexity grows exponentially. The derivation and performance analysis is presented as follows.

8.1.1 Nonlinear Gradient-Based Adaptation Algorithm

Assume that the input signal x(n) and plant noise $n_s(n)$ are both independent zero-mean Gaussian white noise with variances σ_x^2 and $\sigma_{n_s}^2$ respectively. Let the unknown plant be a third-order, M-sample memory truncated Volterra system. The input vector **X**(n) is given by

$$\mathbf{X}(n)=[\underbrace{x(n),\ldots,x(n-M+1)}_{M's},\ \underbrace{x^2(n),\ldots,x^2(n-M+1)}_{M's},$$
$$\underbrace{x(n)x(n-1),\ldots,x(n-M+2)x(n-M+1)}_{\frac{M(M-1)}{2}'s},$$

$$\underbrace{x^3(n),\ldots,x^3(n-M+1)}_{M's},\ \underbrace{x(n)x^2(n-1),\ldots,x^2(n-M+2)x(n-M+1)}_{M(M-1)'s},$$
$$\underbrace{x(n)x(n-1)x(n-2),\ldots,x(n-M+3)x(n-M+2)x(n-M+1)}_{C(M,3)'s}]^T \tag{8.1}$$

The length of vector **X**(n) is L=3M(M+1)/2+C(M,3), where C(M,3) = M!/[3!(M-3)!]. Note that C(M, 3) = 0 if M<3. The weight vector of the unknown Volterra plant corresponding to the Volterra input vector **X**(n) in equation 8.1 is $[w_0^*, w_1^*, ..., w_L^*]^T$. The unknown plant output d(n) is equal to:

$$d(n) = h_0 + \sum_{k_1=0}^{M-1} h_1(k_1)x(n-k_1) + \sum_{k_1=0}^{M-1}\sum_{k_2=0}^{M-1} h_2(k_1,k_2)x(n-k_1)x(n-k_2)$$

$$+\sum_{k_1=0}^{M-1}\sum_{k_2=0}^{M-1}\sum_{k_3=0}^{M-1} h_3(k_1,k_2,k_3)x(n-k_1)x(n-k_2)x(n-k_3) \quad (8.2)$$

As usual, assume the set of *j*th-order Volterra kernel coefficients $\{h_j(k_1,...,k_j),\ 0\le j\le 3\}$ is symmetric. For ease of later use, equation 8.2 can be rewritten as a vector form:

$$d(n) = w_0^* + \mathbf{W}^{*T}\mathbf{X}(n) = \mathbf{C}^{*T}[1, \mathbf{X}^T(n)]^T = \mathbf{C}^{*T}\mathbf{Q}(n) \quad (8.3)$$

where $\mathbf{W}^* = [w_1^*, ..., w_L^*]^T$, $\mathbf{C}^* = [w_0^*, \mathbf{W}^{*T}]^T$, and $\mathbf{Q}(n) = [1, \mathbf{X}^T(n)]^T$. Based on the third-order discrete nonlinear Wiener structure illustrated in figure 8-2, block A contains M-1 delay elements. Four Hermite polynomials, $H_0[z] = 1$, $H_1[z] = z$, $H_2[z] = z^2 - \sigma_x^2$, and $H_3[z] = z^3 - 3z\sigma_x^2$, are used in block B.

Recall from equation 3.82, the input vector $\widetilde{\mathbf{X}}(n)$ of block C is given by

$$\widetilde{\mathbf{X}}(n) = [\widetilde{Q}_0^{(1)}(n), ..., \widetilde{Q}_{M-1}^{(1)}(n), \widetilde{Q}_{00}^{(2)}(n), ..., \widetilde{Q}_{M-1,M-1}^{(2)}(n), \widetilde{Q}_{01}^{(2)}(n), ..., \widetilde{Q}_{M-2,M-1}^{(2)}(n),$$

$$\widetilde{Q}_{000}^{(3)}(n), ..., \widetilde{Q}_{M-1,M-1,M-1}^{(3)}(n), \widetilde{Q}_{011}^{(3)}(n), ..., \widetilde{Q}_{M-2,M-2,M-1}^{(3)}(n)\ \widetilde{Q}_{012}^{(3)}(n), ..., \widetilde{Q}_{M-3,M-2,M-1}^{(3)}(n)\ ]^T$$

$$= [x(n), ..., x(n-M+1), x^2(n)-\sigma_x^2, ..., x^2(n-M+1)-\sigma_x^2, x(n)x(n-1), ..., x(n-M+2)x(n-M+1),$$

$$x^3(n)-3x(n), ..., x^3(n-M+1)-3x(n-M+1), x(n)(x^2(n-1)-\sigma_x^2), .., (x^2(n-M+2)-\sigma_x^2)x(n-M+1),$$

$$x(n)x(n-1)x(n-2), ..., x(n-M+3)x(n-M+2)x(n-M+1)\ ]^T \quad (8.4)$$

The vector length of $\widetilde{\mathbf{X}}(n)$ is the same as **X**(n). The output of the adaptive plant is

$$y(n) = c_0 + \sum_{n_1=0}^{M-1} c_1(n_1)\widetilde{Q}^{(1)}_{n_1}(n) + \sum_{n_1=0}^{M-1}\sum_{n_2=0}^{M-1} c_2(n_1,n_2)\widetilde{Q}^{(2)}_{n_1n_2}(n)$$
$$+\sum_{n_1=0}^{M-1}\sum_{n_2=0}^{M-1}\sum_{n_3=0}^{M-1} c_3(n_1,n_2,n_3)\widetilde{Q}^{(3)}_{n_1n_2n_3}(n) \quad (8.5)$$

where $\{c_j(n_1,...,n_j),\ 0\le j\le 3\}$ is the set of *j*th-order nonlinear Wiener kernel coefficients. Corresponding to the nonlinear Wiener kernel coefficients in equation 8.5, define the weight-vector of the Wiener adaptive system as $[w_0(n), w_1(n), ..., w_L(n)]^T$. Thus, equation 8.5 can be rewritten as

$$y(n) = w_0(n) + \mathbf{W}^T(n)\,\mathbf{S}^{-1}_{\widetilde{X}}\widetilde{\mathbf{X}}(n) = \mathbf{C}^T(n)\,\mathbf{S}^{-1}_{\widetilde{Q}}[1, \widetilde{\mathbf{X}}^T(n)]^T = \mathbf{C}^T(n)\,\mathbf{S}^{-1}_{\widetilde{Q}}\widetilde{\mathbf{Q}}(n) \quad (8.6)$$

where $\mathbf{W}(n) = [w_1(n), ..., w_L(n)]^T$, $\mathbf{C}(n) = [w_0(n), \mathbf{W}^T(n)]^T$, and $\widetilde{\mathbf{Q}}(n) = [1, \widetilde{\mathbf{X}}^T(n)]^T$. $\mathbf{S}^{-1}_{\widetilde{X}}$ is a diagonal scalar matrix which is defined as:

$$\mathbf{S}_{\widetilde{X}} = \mathbf{R}_{\widetilde{X}\widetilde{X}}^{1/2} \quad (8.7)$$

where $\mathbf{R}_{\widetilde{X}\widetilde{X}} = E\{\widetilde{\mathbf{X}}(n)\widetilde{\mathbf{X}}^T(n)\}$. It implies that $\mathbf{S}^{-1}_{\widetilde{X}}\mathbf{R}_{\widetilde{X}\widetilde{X}}\mathbf{S}^{-1}_{\widetilde{X}}$ is an identity matrix. The error signal in figure 8-1 can be expressed as

$$e(n) = d(n) - y(n) + n_s(n) = w_0^* - w_0(n) + \mathbf{W}^{*T}\mathbf{X}(n) - \mathbf{W}^T(n)\,\mathbf{S}^{-1}_{\widetilde{X}}\widetilde{\mathbf{X}}(n) + n_s(n) \quad (8.8)$$

In order to find the optimal weight vector, assume that $\mathbf{C} = [w_0, \mathbf{W}^T]^T$, which does not change with time. Therefore, the mean-square error $\xi = E\{e^2(n)\}$ is obtained by expanding equation 8.8:

$$\xi = E\{n_s^2(n)\} + (w_0^* - w_0)^2 + 2w_0^*E\{\mathbf{X}^T(n)\}\mathbf{W}^* - 2w_0E\{\mathbf{X}^T(n)\}\mathbf{W}^*$$
$$+ \mathbf{W}^{*T}\mathbf{R}_{XX}\mathbf{W}^* + \mathbf{W}^T\mathbf{S}^{-1}_{\widetilde{X}}\mathbf{R}_{\widetilde{X}\widetilde{X}}\mathbf{S}^{-1}_{\widetilde{X}}\mathbf{W} - 2\mathbf{W}^T\mathbf{S}^{-1}_{\widetilde{X}}\mathbf{R}_{\widetilde{X}X}\mathbf{W}^* \quad (8.9)$$

where $\mathbf{R}_{XX} = E\{\mathbf{X}(n)\mathbf{X}^T(n)\}$ and $\mathbf{R}_{\widetilde{X}X} = E\{\widetilde{\mathbf{X}}(n)\mathbf{X}^T(n)\}$. It is worth noting that the relationship of the three correlation matrices $\mathbf{R}_{XX}$, $\mathbf{R}_{\widetilde{X}\widetilde{X}}$ and $\mathbf{R}_{\widetilde{X}X}$ is (see appendix 8A for the derivation of this relationship):

$$\boxed{\mathbf{R}_{XX} = \mathbf{R}^T_{\widetilde{X}X}\mathbf{R}^{-T}_{\widetilde{X}\widetilde{X}}\mathbf{R}_{\widetilde{X}X} + \mathbf{M}_x\mathbf{M}_x^T} \quad (8.10)$$

where $\mathbf{M}_x = E\{\mathbf{X}(n)\}$. To minimize the mean-square error, taking the true gradient of equation 8.9 with respect to $\mathbf{C} = [w_0, \mathbf{W}^T]^T$ yields:

$$\nabla\xi = \begin{bmatrix} \partial\xi/\partial w_0 \\ \partial\xi/\partial \mathbf{W} \end{bmatrix} = \begin{bmatrix} 2w_0 - 2w_0^* - 2E\{\mathbf{X}^T(n)\}\mathbf{W}^* \\ 2\mathbf{S}_{\tilde{x}}^{-1}\mathbf{R}_{\tilde{X}\tilde{X}}\mathbf{S}_{\tilde{x}}^{-1}\mathbf{W} - 2\mathbf{S}_{\tilde{x}}^{-1}\mathbf{R}_{\tilde{X}X}\mathbf{W}^* \end{bmatrix} \tag{8.11}$$

where ∇ is the gradient operator. Setting $\nabla\xi = 0$, the optimal solution of the weight-vector is obtained as:

$$\boxed{\mathbf{C}_{optm} = \begin{bmatrix} w_{0/optm} \\ \mathbf{W}_{optm} \end{bmatrix} = \begin{bmatrix} 1 & E\{\mathbf{X}^T(n)\} \\ \mathbf{0} & \mathbf{S}_{\tilde{x}}\mathbf{R}_{\tilde{X}\tilde{X}}^{-1}\mathbf{R}_{\tilde{X}X} \end{bmatrix} \begin{bmatrix} w_0^* \\ \mathbf{W}^* \end{bmatrix}} \tag{8.12}$$

Substituting equation 8.12 in equation 8.9 gives the minimum mean-square error

$$\xi_{min} = E\{n_s^2(n)\} - (E\{\mathbf{X}^T(n)\}\mathbf{W}^*)^2 + \mathbf{W}^{*T}\mathbf{R_{XX}}\mathbf{W}^* - \mathbf{W}^{*T}\mathbf{R}_{\tilde{X}X}^T\mathbf{R}_{\tilde{X}\tilde{X}}^{-T}\mathbf{R}_{\tilde{X}X}\mathbf{W}^* \tag{8.13}$$

From equation 8.10, ξ_{min} can be expressed simply as:

$$\boxed{\xi_{min} = E\{n_s^2(n)\} = \sigma_{n_s}^2} \tag{8.14}$$

In order to show that the mean-square error $\xi(n)$ can be expressed in the quadratic form of $\mathbf{C}_{optm}$, let $\mathbf{S}_{\tilde{Q}}^{-1} = \begin{bmatrix} 1 & \mathbf{0}^T \\ \mathbf{0} & \mathbf{S}_{\tilde{X}}^{-1} \end{bmatrix}$ and define the weight deviation vector $\mathbf{V}(n)$ as follows:

$$\mathbf{V}(n) = \mathbf{S}_{\tilde{Q}}^{-1}(\mathbf{C}(n) - \mathbf{C}_{optm}) = \mathbf{S}_{\tilde{Q}}^{-1}\mathbf{C}(n) - \begin{bmatrix} 1 & \mathbf{0}^T \\ \mathbf{0} & \mathbf{R}_{\tilde{X}\tilde{X}}^{-1}\mathbf{R}_{\tilde{X}X} \end{bmatrix}\mathbf{C}^* - \begin{bmatrix} E\{\mathbf{X}^T(n)\}\mathbf{W}^* \\ \mathbf{0} \end{bmatrix} \tag{8.15}$$

where $\mathbf{0}$ is a zero column vector. Therefore, equation 8.9 may be rewritten in terms of $\mathbf{R}_{\tilde{Q}\tilde{Q}} = \begin{bmatrix} 1 & \mathbf{0}^T \\ \mathbf{0} & \mathbf{R}_{\tilde{X}\tilde{X}} \end{bmatrix}$ and $\mathbf{V}(n)$, which is (see appendix 8B for the derivation):

$$\xi(n) = \xi_{min} + \mathbf{V}^T(n)\mathbf{R}_{\tilde{Q}\tilde{Q}}\mathbf{V}(n) \tag{8.16}$$

Differentiating equation 8.16 with respect to $\mathbf{C}(n)$ leads to:

$$\frac{\partial\xi(n)}{\partial\mathbf{C}(n)} = 2\mathbf{S}_{\tilde{Q}}^{-1}\mathbf{R}_{\tilde{Q}\tilde{Q}}\mathbf{V}(n) \tag{8.17}$$

Equations 8.16 and 8.17 will be used to develop the performance analysis of the nonlinear Wiener LMS algorithm in the next subsection. From the derivations shown above, we note that equation 8.12 provides an optimal solution such that the minimum mean-square error in equation 8.14 can be achieved. However, this result is hard to implement in real time and suffers from computational difficulties, because we do not know the unknown plant's weight vector and the statistical characteristics of input autocorrelation matrices in advance. A solution is to develop the nonlinear Wiener LMS algorithm which is presented as follows.

8.1.2 Third-Order Nonlinear Wiener LMS Algorithm and Performance Analysis

To derive the third-order nonlinear Wiener-model-based LMS algorithm, we need to consider the instantaneous version of the error power. Define $\varepsilon(n)$ as $e^2(n)$ which is given by:

$$\begin{aligned}\varepsilon(n) = {} & [n_s(n) + w_0^* - w_0(n)]^2 + \mathbf{W}^{*T}\mathbf{X}(n)\mathbf{X}^T(n)\,\mathbf{W}^* \\ & +\mathbf{W}^T(n)\,\mathbf{S}_{\tilde{\mathbf{X}}}^{-1}\,\tilde{\mathbf{X}}(n)\,\tilde{\mathbf{X}}^T(n)\,\mathbf{S}_{\tilde{\mathbf{X}}}^{-1}\,\mathbf{W}(n) \\ & +2n_s(n)\,\mathbf{X}^T(n)\mathbf{W}^* + 2w_0^*\,\mathbf{X}^T(n)\mathbf{W}^* - 2w_0(n)\mathbf{X}^T(n)\mathbf{W}^* \\ & -2n_s(n)\,\tilde{\mathbf{X}}^T(n)\,\mathbf{S}_{\tilde{\mathbf{X}}}^{-1}\mathbf{W}(n) \\ & -2w_0^*\tilde{\mathbf{X}}^T(n)\mathbf{S}_{\tilde{\mathbf{X}}}^{-1}\mathbf{W}(n) - 2w_0(n)\,\tilde{\mathbf{X}}^T(n)\mathbf{S}_{\tilde{\mathbf{X}}}^{-1}\mathbf{W}(n) \\ & -2\mathbf{W}^T(n)\,\mathbf{S}_{\tilde{\mathbf{X}}}^{-1}\,\tilde{\mathbf{X}}(n)\,\mathbf{X}^T(n)\mathbf{W}^*\end{aligned} \tag{8.18}$$

Differentiating equation 8.18 with respect to $\mathbf{C}(n)$, we have

$$\begin{aligned}\nabla\varepsilon(n) &= \begin{bmatrix}\partial\varepsilon(n)/\partial w_0(n) \\ \partial\varepsilon(n)/\partial\mathbf{W}(n)\end{bmatrix} \\ &= \begin{bmatrix}-2[n_s(n)+w_0^*+\mathbf{X}^T(n)\mathbf{W}^* - w_0(n)-\tilde{\mathbf{X}}^T(n)\mathbf{S}_{\tilde{\mathbf{X}}}^{-1}\mathbf{W}(n)] \\ -2\mathbf{S}_{\tilde{\mathbf{X}}}^{-1}\tilde{\mathbf{X}}^T(n)[n_s(n)+w_0^*+\mathbf{X}^T(n)\mathbf{W}^* - w_0(n)-\tilde{\mathbf{X}}^T(n)\mathbf{S}_{\tilde{\mathbf{X}}}^{-1}\mathbf{W}(n)]\end{bmatrix} \\ &= \begin{bmatrix}-2e(n) \\ -2\mathbf{S}_{\tilde{\mathbf{X}}}^{-1}\tilde{\mathbf{X}}^T(n)e(n)\end{bmatrix} = -2e(n)\mathbf{S}_{\tilde{\mathbf{Q}}}^{-1}\,\tilde{\mathbf{Q}}(n)\end{aligned} \tag{8.19}$$

As in the method of steepest descent (see chapter 5), the weight-update equation is expressed as (dropping the ^ hat over C):

$$\mathbf{C}(n+1) = \mathbf{C}(n) - \mu\nabla\varepsilon(n) \tag{8.20}$$

where μ is the step size. Substituting equation 8.20 into equation 8.18 leads to:

$$\boxed{\mathbf{C}(n+1) = \mathbf{C}(n) + 2\mu e(n)\mathbf{S}_{\tilde{\mathbf{Q}}}^{-1}\tilde{\mathbf{Q}}(n)} \tag{8.21}$$

Equation 8.21 is the weight-update equation of the discrete nonlinear Wiener LMS algorithm. Note that in equation 8.21, the current weight vector is a function of the past input vectors only. Based on the independence assumption (Haykin 1996), we can assume $\mathbf{C}(n)$ is independent of $\tilde{\mathbf{Q}}(n)$.

In order to examine the convergence of equation 8.21 and derive the step-size range, taking the expectation of equation 8.21 on both sides, $E\{\mathbf{C}(n+1)\}$ can be expressed in terms of $\mathbf{R}_{\tilde{Q}\tilde{Q}}$ and $\mathbf{R}_{\tilde{Q}Q} = \begin{bmatrix} 1 & \mathbf{0}^T \\ \mathbf{0} & \mathbf{R}_{\tilde{X}X} \end{bmatrix}$, which is:

$$E\{\mathbf{C}(n+1)\} = (\mathbf{I} - 2\mu\mathbf{S}_{\tilde{Q}}^{-2}\mathbf{R}_{\tilde{Q}\tilde{Q}})E\{\mathbf{C}(n)\} + 2\mu\mathbf{S}_{\tilde{Q}}^{-1}\mathbf{R}_{\tilde{Q}Q}\mathbf{C}^* + 2\mu\begin{bmatrix} E\{\mathbf{X}^T(n)\}\mathbf{W}^* \\ \mathbf{0} \end{bmatrix} \tag{8.22}$$

where $\mathbf{I}$ is an identity matrix. To see this, substituting equation 8.19 in equation 8.20 and taking the expectation on both sides, and using the assumption that $\mathbf{C}(n)$ and $\tilde{\mathbf{Q}}(n)$ are independent, we have:

$$E\{\begin{bmatrix} w_0(n+1) \\ \mathbf{W}(n+1) \end{bmatrix}\} = E\{\begin{bmatrix} w_0(n) - \mu\partial\varepsilon(n)/\partial w_0(n) \\ \mathbf{W}(n) - \mu\partial\varepsilon(n)/\partial\mathbf{W}(n) \end{bmatrix}\}$$

$$= E\{\begin{bmatrix} w_0(n) \\ \mathbf{W}(n) \end{bmatrix}\} - 2\mu\begin{bmatrix} 1 & \mathbf{0}^T \\ \mathbf{0} & \mathbf{S}_{\tilde{X}}^{-2} \end{bmatrix}\begin{bmatrix} 1 & \mathbf{0}^T \\ \mathbf{0} & E\{\tilde{\mathbf{X}}^T(n)\tilde{\mathbf{X}}(n)\} \end{bmatrix} E\{\begin{bmatrix} w_0(n) \\ \mathbf{W}(n) \end{bmatrix}\}$$

$$+2\mu\begin{bmatrix} 1 & \mathbf{0}^T \\ \mathbf{0} & \mathbf{S}_{\tilde{X}}^{-1} \end{bmatrix}\begin{bmatrix} 1 & E\{\mathbf{X}^T(n)\} \\ \mathbf{0} & E\{\tilde{\mathbf{X}}^T(n)\mathbf{X}(n)\} \end{bmatrix}\begin{bmatrix} w_0^* \\ \mathbf{W}^* \end{bmatrix}$$

$$= (\mathbf{I} - 2\mu\begin{bmatrix} 1 & \mathbf{0}^T \\ \mathbf{0} & \mathbf{S}_{\tilde{X}}^{-2} \end{bmatrix}\begin{bmatrix} 1 & \mathbf{0}^T \\ \mathbf{0} & \mathbf{R}_{\tilde{X}\tilde{X}} \end{bmatrix})E\{\begin{bmatrix} w_0(n) \\ \mathbf{W}(n) \end{bmatrix}\} +$$

$$2\mu\begin{bmatrix} 1 & \mathbf{0}^T \\ \mathbf{0} & \mathbf{S}_{\tilde{X}}^{-1} \end{bmatrix}\begin{bmatrix} 1 & E\{\mathbf{X}^T(n)\} \\ \mathbf{0} & \mathbf{R}_{\tilde{X}X} \end{bmatrix}\begin{bmatrix} w_0^* \\ \mathbf{W}^* \end{bmatrix}$$

$$= (\mathbf{I} - 2\mu\mathbf{S}_{\tilde{Q}}^{-2}\mathbf{R}_{\tilde{Q}\tilde{Q}})E\{\mathbf{C}(n)\} + 2\mu\mathbf{S}_{\tilde{Q}}^{-1}\mathbf{R}_{\tilde{Q}Q}\mathbf{C}^* + 2\mu\begin{bmatrix} E\{\mathbf{X}^T(n)\}\mathbf{W}^* \\ \mathbf{0} \end{bmatrix}$$

According to the definition of the weight deviation vector in equation 8.15, E{**C**(n)} can be found as:

$$\mathrm{E}\{\mathbf{C}(\mathrm{n})\} = \mathbf{S}_{\tilde{\mathrm{Q}}}\left(\mathrm{E}\{\mathbf{V}(\mathrm{n})\} + \begin{bmatrix} 1 & \mathbf{0}^{\mathrm{T}} \\ \mathbf{0} & \mathbf{R}_{\tilde{\mathbf{X}}\tilde{\mathbf{X}}}^{-1}\mathbf{R}_{\tilde{\mathbf{X}}\mathbf{X}} \end{bmatrix} \mathbf{C}^{*} + \begin{bmatrix} \mathrm{E}\{\mathbf{X}^{\mathrm{T}}(\mathrm{n})\}\mathbf{W}^{*} \\ \mathbf{0} \end{bmatrix}\right) \tag{8.23}$$

Substituting equation 8.23 into equation 8.22, direct calculation yields the recursive expression for E{**V**(n)}:

$$\mathrm{E}\{\mathbf{V}(\mathrm{n+1})\} = (\mathbf{I} - 2\,\mu \mathbf{S}_{\tilde{\mathrm{Q}}}^{-2}\mathbf{R}_{\tilde{\mathrm{Q}}\tilde{\mathrm{Q}}})\mathrm{E}\{\mathbf{V}(\mathrm{n})\} \tag{8.24}$$

Recall that $\mathbf{R}_{\tilde{\mathrm{Q}}\tilde{\mathrm{Q}}} = \tilde{\mathbf{P}}\,\tilde{\mathbf{D}}\,\tilde{\mathbf{P}}^{-1}$. Then, replacing **V**(n) by $\tilde{\mathbf{P}}\,\mathbf{V}'(\mathrm{n})$ in equation 8.24, pre-multiplying by $\tilde{\mathbf{P}}^{-1}$ on both sides yields:

$$\mathrm{E}\{\mathbf{V}'(\mathrm{n})\} = (\mathrm{I} - 2\mu \mathbf{S}_{\tilde{\mathrm{Q}}}^{-2}\mathbf{D})^{\mathrm{n}}\mathbf{V}'(0) \tag{8.25}$$

where **V**′(**0**) is the initial vector. For convergence, the range of step-size should be:

$$0 < \mu < \frac{1}{\lambda_{\max}} \quad \text{or} \quad 0 < \mu < \frac{1}{\mathrm{tr}[\mathbf{S}_{\tilde{\mathrm{Q}}}^{-2}\mathbf{R}_{\tilde{\mathrm{Q}}\tilde{\mathrm{Q}}}]} < \frac{1}{\lambda_{\max}} \tag{8.26}$$

where $\lambda_{\max}$ is the maximum eigenvalue of $\mathbf{S}_{\tilde{\mathrm{Q}}}^{-2}\mathbf{R}_{\tilde{\mathrm{Q}}\tilde{\mathrm{Q}}}$. Because $\mathbf{S}_{\tilde{\mathrm{Q}}}^{-2}\mathbf{R}_{\tilde{\mathrm{Q}}\tilde{\mathrm{Q}}}$ is an identity matrix, this implies $\lambda_{\max} = 1$ and $\mathrm{tr}[\mathbf{S}_{\tilde{\mathrm{Q}}}^{-2}\mathbf{R}_{\tilde{\mathrm{Q}}\tilde{\mathrm{Q}}}]=\mathrm{L}+1$. Then equation 8.26 becomes:

$$0 < \mu < 1 \quad \text{or} \quad 0 < \mu < \frac{1}{\mathrm{L}+1} < 1 \tag{8.27}$$

Equation 8.27 is the condition that guarantees that the expected weight vector can reach the optimum solution. This means, if equation 8.26 is satisfied , it follows that:

$$\lim_{\mathrm{n}\to\infty} \mathrm{E}\{\mathbf{V}'(\mathrm{n})\} = \lim_{\mathrm{n}\to\infty} \mathrm{E}\{\tilde{\mathbf{P}}^{-1}\mathbf{V}(\mathrm{n})\} = 0 \tag{8.28}$$

Recall from equation 8.27, we can find that the expected value of **C**(n) is

$$\lim_{\mathrm{n}\to\infty} \mathrm{E}\{\mathbf{C}(\mathrm{n})\} = \mathbf{C}_{\mathrm{optm}} \tag{8.29}$$

Therefore, the nonlinear Wiener steepest descent and LMS algorithm are generally stable and convergent if the condition in equation 8.26 is met.

To derive the misadjustment, we need to consider the steady state condition. In steady state , equation 8.9 can be expressed as:

$$\hat{\xi}(n) = \sigma_{n_s}^2 + w_0^{*2} + \hat{w}_0^2(n) - 2w_0^*\ \hat{w}_0(n) + 2w_0^* E\{\mathbf{X}^T(n)\}\mathbf{W}^* - 2\hat{w}_0(n)E\{\mathbf{X}^T(n)\}\mathbf{W}^*$$
$$+\mathbf{W}^{*T}\mathbf{R}_{\mathbf{XX}}\mathbf{W}^* + \hat{\mathbf{W}}^T(n)\,\mathbf{S}_{\tilde{\mathbf{X}}}^{-1}\,\mathbf{R}_{\tilde{\mathbf{X}}\tilde{\mathbf{X}}}\mathbf{S}_{\tilde{\mathbf{X}}}^{-1}\hat{\mathbf{W}}(n) - 2\hat{\mathbf{W}}^T(n)\,\mathbf{S}_{\tilde{\mathbf{X}}}^{-1}\,\mathbf{R}_{\tilde{\mathbf{X}}\mathbf{X}}\mathbf{W}^* \tag{8.30}$$

where the header ^ means steady state. In steady state, from equation 8.13 we can assume that:

$$\hat{w}_0(n) \approx w_0^* + E\{\mathbf{X}^T(n)\}\mathbf{W}^* \tag{8.31a}$$

$$\mathbf{S}_{\tilde{\mathbf{X}}}^{-1}\hat{\mathbf{W}}(n) \approx \mathbf{R}_{\tilde{\mathbf{X}}\tilde{\mathbf{X}}}^{-1}\mathbf{R}_{\tilde{\mathbf{X}}\mathbf{X}}\mathbf{W}^* \tag{8.31b}$$

Substituting equations 8.31a and 8.31b in 8.30, using the relationship $\mathbf{R}_{\tilde{\mathbf{X}}\tilde{\mathbf{X}}} = (\mathbf{R}_{\tilde{\mathbf{X}}\mathbf{X}}^{-1}\mathbf{R}_{\tilde{\mathbf{X}}\tilde{\mathbf{X}}})^T(\mathbf{R}_{\mathbf{XX}} - \mathbf{M}_x\mathbf{M}_x^T)\ \mathbf{R}_{\tilde{\mathbf{X}}\mathbf{X}}^{-1}\mathbf{R}_{\tilde{\mathbf{X}}\tilde{\mathbf{X}}}$ (see appendix 8A), proper manipulating and rearranging leads to:

$$\hat{\xi}(n) = \xi_{\min} - (E\{\mathbf{X}^T(n)\}\mathbf{W})^2 + \mathbf{W}^{*T}(\mathbf{M}_x\mathbf{M}_x^T + \mathbf{R}_{\mathbf{XX}} - \mathbf{M}_x\mathbf{M}_x^T)\mathbf{W}^* - \hat{\mathbf{W}}^T(n)\,\mathbf{S}_{\tilde{\mathbf{X}}}^{-1}\,\mathbf{R}_{\tilde{\mathbf{X}}\tilde{\mathbf{X}}}\mathbf{S}_{\tilde{\mathbf{X}}}^{-1}\hat{\mathbf{W}}(n)$$

$$\approx \xi_{\min} + [(\mathbf{R}_{\tilde{\mathbf{X}}\mathbf{X}}^{-1}\mathbf{R}_{\tilde{\mathbf{X}}\tilde{\mathbf{X}}}\mathbf{W}^*)^T - \mathbf{S}_{\tilde{\mathbf{X}}}^{-1}\hat{\mathbf{W}}^T(n)]\mathbf{R}_{\tilde{\mathbf{X}}\tilde{\mathbf{X}}}[\mathbf{R}_{\tilde{\mathbf{X}}\mathbf{X}}^{-1}\mathbf{R}_{\tilde{\mathbf{X}}\tilde{\mathbf{X}}}\mathbf{W}^* - \mathbf{S}_{\tilde{\mathbf{X}}}^{-1}\hat{\mathbf{W}}(n)]$$

$$\approx \xi_{\min} + \hat{\mathbf{V}}^T(n)\,\mathbf{R}_{\tilde{\mathbf{Q}}\tilde{\mathbf{Q}}}\hat{\mathbf{V}}(n) \tag{8.32}$$

where $\hat{\mathbf{V}}(n) = \mathbf{S}_{\tilde{\mathbf{Q}}}^{-1}(\hat{\mathbf{C}}(n) - \mathbf{C}_{\text{optm}}) = \mathbf{S}_{\tilde{\mathbf{Q}}}^{-1}\hat{\mathbf{C}}(n) - \begin{bmatrix} 1 & \mathbf{0}^T \\ \mathbf{0} & \mathbf{R}_{\tilde{\mathbf{X}}\tilde{\mathbf{X}}}^{-1}\mathbf{R}_{\tilde{\mathbf{X}}\mathbf{X}} \end{bmatrix}\mathbf{C}^* - \begin{bmatrix} E\{\mathbf{X}^T(n)\}\mathbf{W}^* \\ \mathbf{0} \end{bmatrix}$. Note that equation 8.32 is a steady state result of equation 8.16. Define the excess mean-square error as:

$$\text{excess MSE} = E\{\hat{\xi}(n) - \xi_{\min}\} = E\{\hat{\mathbf{V}}^T(n)\,\mathbf{R}_{\tilde{\mathbf{Q}}\tilde{\mathbf{Q}}}\hat{\mathbf{V}}(n)\}$$

$$= E\{\hat{\mathbf{V}}^T(n)(\tilde{\mathbf{P}}^T\tilde{\mathbf{D}}\tilde{\mathbf{P}})\,\hat{\mathbf{V}}(n)\} = E\{\hat{\mathbf{V}}'^T(n)\,\tilde{\mathbf{D}}\,\hat{\mathbf{V}}'(n)\} \tag{8.33}$$

where $\hat{\mathbf{V}}'(n) = \tilde{\mathbf{P}}^T\,\hat{\mathbf{V}}(n)$. Now consider the steady state error gradient which is defined as:

$$\hat{\nabla}\varepsilon(n) = \nabla\xi(n) + \mathbf{N}_s(n) \tag{8.34}$$

where $\mathbf{N}_s(n)$ is the gradient estimation noise vector. In steady state, $\nabla\xi(n) \approx 0$. From equation 8.18, the instantaneous steady state error gradient can be obtained as:

$$\hat{\nabla}\varepsilon(n) = -2\,\hat{e}(n)\,\mathbf{S}_{\tilde{Q}}^{-1}\,\tilde{\mathbf{Q}}(n) \approx \mathbf{N}_s(n) \tag{8.35}$$

Define the covariance matrix of $\hat{\mathbf{V}}'(n)$ as $\mathrm{cov}\{\hat{\mathbf{V}}'(n)\} = \mathrm{E}\{\hat{\mathbf{V}}'(n)\,\hat{\mathbf{V}}'^{T}(n)\}$.

To calculate $\mathrm{cov}\{\hat{\mathbf{V}}'(n)\}$, we first assume $\hat{e}(n)$ and $\tilde{\mathbf{Q}}(n)$ are independent. Then the covariance matrix of $\mathbf{N}_s(n)$ can be found approximately as:

$$\mathrm{cov}\{\mathbf{N}_s(n)\} = 4\,\mathrm{E}\{\hat{e}^2(n)\}\,\mathbf{S}_{\tilde{Q}}^{-1}\,\mathrm{E}\{\tilde{\mathbf{Q}}(n)\,\tilde{\mathbf{Q}}^{T}(n)\}\,\mathbf{S}_{\tilde{Q}}^{-1} \approx 4\xi_{\min}\,\mathbf{S}_{\tilde{Q}}^{-1}\,\mathbf{R}_{\tilde{Q}\tilde{Q}}\,\mathbf{S}_{\tilde{Q}}^{-1} \tag{8.36}$$

For the second step, define $\mathbf{N}_s'(n) = \tilde{\mathbf{P}}^{-1}\mathbf{N}_s(n)$. The covariance matrix of $\mathbf{N}_s'$ is obtained as:

$$\mathrm{cov}\{\mathbf{N}_s'(n)\} = \mathrm{cov}\{\tilde{\mathbf{P}}^{-1}\mathbf{N}_s(n)\} = \tilde{\mathbf{P}}^{-1}\,\mathrm{E}\{\mathbf{N}_s(n)\mathbf{N}_s^{T}(n)\}\,\tilde{\mathbf{P}} = 4\xi_{\min}\,\mathbf{S}_{\tilde{Q}}^{-1}\,\tilde{\mathbf{D}}\,\mathbf{S}_{\tilde{Q}}^{-1} \tag{8.37}$$

The third step is to find the recursive steady state weight deviation vector $\hat{\mathbf{V}}(n)$. From equation 8.24, we note that the steady state weight updated equation can be written as:

$$\hat{\mathbf{C}}(n+1) = \hat{\mathbf{C}}(n) - \mu\,\hat{\nabla}\varepsilon(n) \tag{8.38}$$

According to equations 8.15, 8.17, and 8.34, in steady state, $\hat{\mathbf{C}}(n)$ and $\hat{\nabla}\varepsilon(n)$ can be expressed as:

$$\hat{\mathbf{C}}(n) = \mathbf{S}_{\tilde{Q}}\left(\hat{\mathbf{V}}(n) + \begin{bmatrix} 1 & \mathbf{0}^{T} \\ \mathbf{0} & \mathbf{R}_{\tilde{X}\tilde{X}}^{-1}\mathbf{R}_{\tilde{X}X} \end{bmatrix}\mathbf{C}^{*} + \begin{bmatrix} \mathrm{E}\{\mathbf{X}^{T}(n)\}\mathbf{W}^{*} \\ \mathbf{0} \end{bmatrix}\right) \tag{8.39}$$

$$\hat{\nabla}\varepsilon(n) = 2\mathbf{S}_{\tilde{Q}}^{-1}\,\mathbf{R}_{\tilde{Q}\tilde{Q}}\,\hat{\mathbf{V}}(n) + \mathbf{N}_s(n) \tag{8.40}$$

Substituting equation 8.39 and equation 8.40 in equation 8.38 and rearranging yields:

$$\hat{\mathbf{V}}'(n+1) = (\mathbf{I} - 2\mu\,\mathbf{S}_{\tilde{Q}}^{-2}\,\tilde{\mathbf{D}})\,\hat{\mathbf{V}}'(n) - \mu\,\mathbf{S}_{\tilde{Q}}^{-1}\,\mathbf{N}_s'(n) \tag{8.41}$$

where $\mathbf{N}'_s(n) = \widetilde{\mathbf{P}}^{-1}\mathbf{N}_s(n)$. From equation 8.41, the covariance matrix of $\hat{\mathbf{V}}'(n)$ is found as:

$$\operatorname{cov}\{\hat{\mathbf{V}}'(n)\} = (\mathbf{I} - 2\mu\mathbf{S}_{\tilde{\mathbf{Q}}}^{-2}\widetilde{\mathbf{D}})E\{\hat{\mathbf{V}}'(n-1)\hat{\mathbf{V}}'^{T}(n-1)\}(\mathbf{I} - 2\mu\mathbf{S}_{\tilde{\mathbf{Q}}}^{-2}\widetilde{\mathbf{D}})^{T}$$

$$+\mu^2\mathbf{S}_{\tilde{\mathbf{Q}}}^{-1}E\{\mathbf{N}'_s(n-1)\mathbf{N}'^{T}_s(n-1)\}\mathbf{S}_{\tilde{\mathbf{Q}}}^{-1} \tag{8.42}$$

Collecting all the $\operatorname{cov}\{\hat{\mathbf{V}}'(n)\}$ terms on the left hand side leads to:

$$\operatorname{cov}\{\hat{\mathbf{V}}'(n)\} = \frac{\mu}{4(\mathbf{S}_{\tilde{\mathbf{Q}}}^{-2}\widetilde{\mathbf{D}} - \mu\mathbf{S}_{\tilde{\mathbf{Q}}}^{-4}\widetilde{\mathbf{D}}^2)}\operatorname{cov}\{\mathbf{N}'_s(n)\} \tag{8.43}$$

Because $\mu\mathbf{S}^{-4}\hat{\mathbf{D}}^{-2} \ll \mathbf{I}$, and from equation 8.37, equation 8.43 can be simplified as:

$$\operatorname{cov}\{\hat{\mathbf{V}}'(n)\} \approx \mu\,\xi_{\min}\,\mathbf{S}_{\tilde{\mathbf{Q}}}^{-2} \tag{8.44}$$

Therefore, the excess mean squared error in equation 8.33 is expressed as:

$$\text{excess MSE} = E\{\hat{\mathbf{V}}'^{T}(n)\widetilde{\mathbf{D}}\hat{\mathbf{V}}'(n)\} \approx \mu\,\xi_{\min}\operatorname{tr}[\mathbf{S}_{\tilde{\mathbf{Q}}}^{-2}\mathbf{R}_{\tilde{\mathbf{Q}}\tilde{\mathbf{Q}}}] \tag{8.45}$$

Finally, the misadjustment can be calculated according to:

$$\boxed{\text{MISADJ} = \frac{\text{excess MSE}}{\xi_{\min}} \approx \mu(L+1)} \tag{8.46}$$

For time constant analysis, let us rewrite equation 8.25 in the scalar form:

$$v'(n) = (1 - 2\mu\lambda_n)^n\, v'(0) \tag{8.47}$$

where λ_n is the eigenvalue of $\mathbf{S}_{\tilde{\mathbf{Q}}}^{-2}\mathbf{D}$. The term $(1-2\mu\lambda_n)^n$ can be approximated by $(e^{1/\tau_n})^n$. If τ_n is large, then

$$1 - 2\mu\lambda_n \approx e^{1/\tau_n} \approx 1 - \frac{1}{\tau_n} \tag{8.48}$$

The time constant can be found as:

$$\boxed{\tau_n \approx \frac{1}{2\mu\lambda_n} = \frac{1}{2\mu}} \tag{8.49}$$

A count of the arithmetic operations involved in the implementation of the algorithm listed in table 8.1 shows that it requires $(5M^3+33M^2+40M+18)/6$ multiplications per iteration. Our approach therefore has $O(M^3)$ computational complexity.

Table 8.1 Computational complexity of the third-order onlinear Wiener LMS filter (M>3)

Initialization:	$\mathbf{C}(n)=\mathbf{0}$ Pre-calculate $\mathbf{S}_{\tilde{\mathbf{Q}}}^{-1}$ defined in equation(8.7) $\mu'=2\mu$		
	Relation	Dimension	Multiplication Count
Block A	$\mathbf{Z}(n)=[x(n), x(n-1), ..., x(n-M+1)]^T$	$M\times 1$	0
Block B	$\tilde{\mathbf{Q}}(n)=[1\ \tilde{\mathbf{X}}(n)\ ]^T$ $\tilde{\mathbf{X}}(n)$ is defined by equation(8.4)	$(L+1)\times 1$	$(2M^3+15M^2+7M)/6$
Block C	$\mathbf{C}(n+1)=\mathbf{C}(n)+\mu' e(n)\mathbf{S}_{\tilde{\mathbf{Q}}}^{-1}\tilde{\mathbf{Q}}(n)$	$(L+1)\times 1$	$3(L+1)$
			Total: $(5M^3+33M^2+40M+18)/6$

Remark: Extension of our algorithm and performance analysis to identify the higher-order Volterra system is straightforward. In fact, for *P*th-order, M-sample memory nonlinear Wiener LMS filter, the vector in block B and C will have $O(M^P)$ elements; consequently, it requires $O(M^P)$ multiplications per iteration.

8.2 Computer Simulation Results

Example 1: Consider a general third-order Volterra filter of 3-sample memory which has the following input-output relationship:

$$\begin{aligned}y(n) = &-0.78x(n) - 1.48x(n-1) + 1.39x(n-2) + 0.76x(n-1)x(n-2)\\ &+ 1.86x(n)x(n-2) + 3.72x(n)x(n-1) + 1.41x^2(x-2) - 1.62x^2(n-1)\\ &+ 0.54x^2(n) + 0.33x(n)x(n-1)x(n-2) + 0.15x^2(n-1)x(n-2)\\ &- 0.75x^2(n-2)x(n-1) - 1.52x^2(n)x(n-2) - 0.23x^2(n-2)x(n)\\ &- 0.12x^2(n)x(n-1) - 0.13x^2(n-1)x(n) + 0.5x^3(n-2) - 0.76x^3(n-1)\\ &+ 0.04x^3(n)\end{aligned} \tag{8.50}$$

We use the delay line structure as in figure 8-2 of length 3 in block A and H_0, H_1, and H_2 in block B. For the third-order nonlinear Wiener model, fully expanding equation 8.5, we have a total of 20 $\tilde{Q}$-polynomials, which means that there are 20 coefficients in block C. Properly select the scaling matrix as: $\mathrm{diag}[1,1,1,1,1,1,1,\frac{1}{\sqrt{2}},\frac{1}{\sqrt{2}},\frac{1}{\sqrt{2}},1,\frac{1}{\sqrt{2}},\frac{1}{\sqrt{2}},\frac{1}{\sqrt{2}},\frac{1}{\sqrt{2}},\frac{1}{\sqrt{2}},\frac{1}{\sqrt{2}},\frac{1}{\sqrt{6}},\frac{1}{\sqrt{6}},\frac{1}{\sqrt{6}}]$ which allows the autocorrelation matrix to become an identity matrix. Note that we can also iteratively determine the elements of this scaling matrix. With SNR = –40 db, the simulation results of ensemble averages over 50 independent runs for both the Wiener model and the Volterra model are shown in figure 8-3. For $\mu = 0.0005$, we see that the Wiener model adaptive filter has much better performance than the Volterra model, which does not converge to the right values. For $\mu = 0.003$, the Wiener model adaptive filter is still very stable and converges to the optimal solution as shown in table 8.2. But for the Volterra model it becomes unstable and fails to identify this third-order Volterra system. This is because all $\tilde{Q}$-polynomials in the nonlinear Wiener model are orthogonal, which reduces the eigenvalue spread, and improves the nonlinear LMS adaptive filter performance.

Table 8.2. Adaptive coefficients in steady state of the third-order Wiener system for $\mu = 0.0005$

	$w_0(n)$	$w_1(n)$	$w_2(n)$	$w_3(n)$	$w_4(n)$	$w_5(n)$	$w_6(n)$	$w_7(n)$
Experimental value	0.329902	1.519939	–4.629939	–1.019670	0.759924	1.860451	3.719984	1.994139
Theoretical value	0.330000	1.520000	–4.630000	–1.020000	0.760000	1.860000	3.720000	1.994041

	$w_8(n)$	$w_9(n)$	$w_{10}(n)$	$w_{11}(n)$	$w_{12}(n)$	$w_{13}(n)$	$w_{14}(n)$	$w_{15}(n)$
Experimental value	–2.291214	0.763668	0.330249	0.212218	–1.060616	–2.149371	–0.324744	–0.169223
Theoretical value	–2.291026	0.763675	0.330000	0.212132	–1.060660	–2.149605	–0.325269	–0.169706

	$w_{16}(n)$	$w_{17}(n)$	$w_{18}(n)$	$w_{19}(n)$
Experimental value	–0.183474	1.224797	–1.861081	0.097975
Theoretical value	–0.183848	1.224745	–1.861612	0.097980

The experimental misadjustment of the Wiener model of $\mu = 0.0005$ is equal to 0.012, which is close to the theoretical value 0.01. To examine the eigenvalue spread characteristics, we need to evaluate the eigenvalue of the autocorrelation matrix. The ensemble average autocorrelation matrix of the last 3,000 iterations with 50 runs for both the Wiener model and the Volterra model are shown in figure 8-4 and figure 8-5. Obviously, the autocorrelation matrix of the Wiener model is fully diagonalized. Refer to table 8.3.

Table 8.3. Eigenvalue spread characteristics of the third-order Wiener system

	λ_{max}		λ_{min}		$\lambda_{max} / \lambda_{min}$	
	Experimental value	Theoretical value	Experimental Value	Theoretical value	Experimental value	Theoretical value
Wiener Model	1.060	1	0.950	1	1.12	1
Volterra Model	17.044	17.121	0.266	0.268	64.136	63.779

We can see that for the Volterra model, the eigenvalue spread is equal to 64.136 (compared with the theoretical value 63.779). For the Wiener model, with the help of the scaling matrix, the eigenvalue spread is 1.12 (compared with the theoretical value 1). The eigenvalue spread is reduced about 64 times.

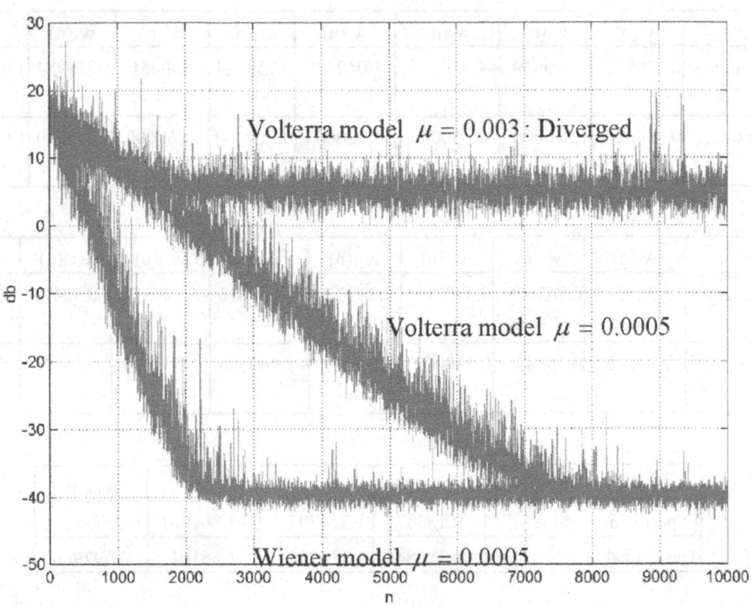

Figure 8-3. Example 1: MSE learning curve of third-order Wiener model and Volterra model

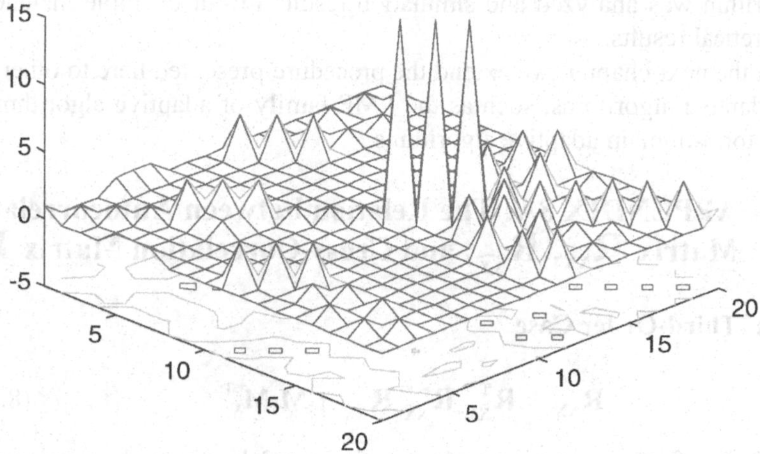

Figure 8-4. Example 1: Autocorrelation matrix of input to third-order Volterra model

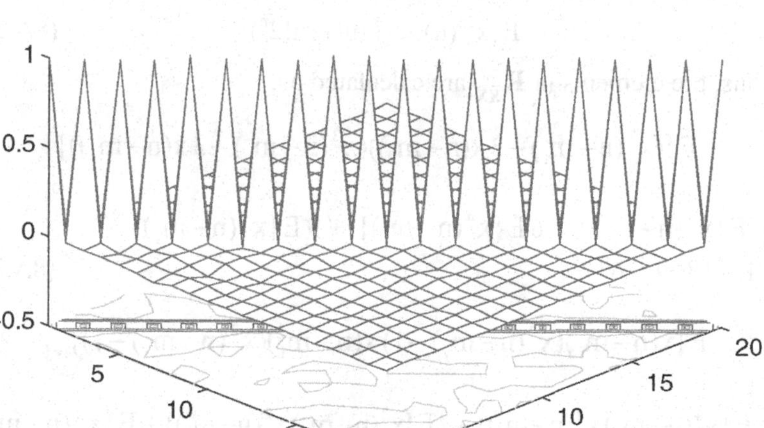

Figure 8-5. Example 1: Autocorrelation matrix of input to third-order Wiener model

8.3 Summary

In this chapter, we developed algorithms based on the third-order nonlinear Wiener models for adaptive system identification. The performance of the algorithm was analyzed and simulation results of our example matched the theoretical results.

In the next chapter, we extend the procedure presented here to other types of adaptive algorithms, such as the LMF family of adaptive algorithms and transform-domain adaptive algorithms.

8.4 APPENDIX 8A: The Relation between Autocorrelation Matrix $\mathbf{R}_{\mathbf{XX}}$, $\mathbf{R}_{\tilde{\mathbf{X}}\tilde{\mathbf{X}}}$, and Cross-Correlation Matrix $\mathbf{R}_{\tilde{\mathbf{X}}\mathbf{X}}$

8.4.1 Third-Order Case

$$\mathbf{R}_{\mathbf{XX}} = \mathbf{R}_{\tilde{\mathbf{X}}\mathbf{X}}^{T}\,\mathbf{R}_{\tilde{\mathbf{X}}\tilde{\mathbf{X}}}^{-T}\,\mathbf{R}_{\tilde{\mathbf{X}}\mathbf{X}} + \mathbf{M}_x\mathbf{M}_x^{T} \tag{8A.2.1}$$

To verify the correctness of equation 8A.2.1, we need to express each correlation matrix in a more explicit manner. For simplicity and without loss of generality, assume the input signal x(n) is Gaussian white noise with unit variance. According to the definitions of the input vectors $\mathbf{X}(n)$ and $\tilde{\mathbf{X}}(n)$ in equation 8.1 and equation 8.4, and according to the high-order joint moment of Gaussian random variables (Papoulis 1991), we know that

$$E\{x^{2n+1}(n)\} = 0 \tag{8A.2.2.a}$$

$$E\{x^{2n}(n)\} = (2n!)/(n!2^{n}) \tag{8A.2.2.b}$$

Thus, the elements in $\mathbf{R}_{\tilde{\mathbf{X}}\tilde{\mathbf{X}}}$ are calculated as:

$$E\{(x^3(n-m_i)-3x(n-m_i))(x^3(n-m_i)-3x(n-m_i))\}$$

$$= E\{x^6(n-m_i)\} - 6E\{x^4(n-m_i)\} + 9E\{x^2(n-m_i)\}$$
$$= 15-18+9 = 6 \tag{8A.2.3.a}$$

$$E\{x(n-m_i)(x^2(n-m_j)-1)\,x(n-m_i)(x^2(n-m_j)-1)\}_{i\neq j}$$

$$= E\{x^4(n-m_j)x^2(n-m_i)\} - 2E\{x^2(n-m_j)x^2(n-m_i)\} + E\{x^2(n-m_i)\}$$

$$= 3-2+1 = 2 \tag{8A.2.3.b}$$

$$E\{x^2(n-m_i)x^2(n-m_j)x^2(n-m_k)\}_{i\neq j\neq k}=1 \tag{8A.2.3.c}$$

Therefore, $\mathbf{R}_{\tilde{X}\tilde{X}}$ can be expressed as:

$$\mathbf{R}_{\tilde{X}\tilde{X}}=\text{diag}[\underbrace{1\ \dots\ 1}_{M's}\ \underbrace{2\ \dots\ 2}_{M's}\ \underbrace{1\ \dots\ 1}_{\frac{M(M-1)}{2}'s}\ \underbrace{6\ \dots\ 6}_{\substack{M,s\\(A.2.3a)}}\underbrace{2\ \dots\ 2}_{\substack{M(M-1),s\\(A.2.3b)}}\underbrace{1\ \dots\ 1}_{\substack{C(M,3),s\\(A.2.3c)}}\ ] \tag{8A.2.4}$$

Note that all the off-diagonal elements involve the odd number terms production of white Gaussian noise whose mean values are zero. Secondarily, due to $E\{x^2(n-m_i)x^2(n-m_j)\}_{i\neq j}=1$, it is not difficult to show that $\mathbf{M}_x$ can be expressed as:

$$\mathbf{M}_x=\text{diag}[\underbrace{0\ \dots\ 0}_{M's}\ \underbrace{1\ \dots\ 1}_{M's}\ \underbrace{0\ \dots\ 0}_{\frac{M(M-1)}{2}'s}\ \underbrace{0\ \dots\ 0}_{M's}\ \underbrace{0\ \dots\ 0}_{M(M-1)'s}\ \underbrace{0\ \dots\ 0}_{C(M,3)'s}] \tag{8A.2.5}$$

Therefore, matrix $\mathbf{M}_x\mathbf{M}_x^T$ can be obtained as:

$$\mathbf{M}_x\mathbf{M}_x^T=\begin{bmatrix}\mathbf{0}_{M\times M} & \mathbf{0} & \mathbf{0}\\ \mathbf{0} & \mathbf{1}_{M\times M} & \mathbf{0}\\ \mathbf{0} & \mathbf{0} & \mathbf{0}_{\frac{M^3+6M^2-M}{6}\times\frac{M^3+6M^2-M}{6}}\end{bmatrix}_{L\times L} \tag{8A.2.6}$$

where $\mathbf{1}_{M\times M}$ is an $M\times M$ matrix with all the elements equal to 1. Third, we note that

$$E\{(x^2(n-m_i)-1)x^2(n-m_i)\}=E\{x^4(n-m_i)\}-E\{x^2(n-m_i)\}=3-1=2 \tag{8A.2.7.a}$$

$$E\{(x^3(n-m_i)-3x(n-m_i))x^3(n-m_i)\}=E\{x^6(n-m_i)\}-3E\{x^4(n-m_i)\}$$
$$=\quad 15-9=6 \tag{8A.2.7.b}$$

$$E\{x(n-m_i)(x^2(n-m_j)-1)\,x(n-m_i)x^2(n-m_j)\}_{i\neq j}=E\{x^2(n-m_i)x^4(n-m_j)\}-E\{x^2(n-m_i)x^2(n-m_j)\}$$

$$=3-1=2 \tag{8A.2.7.c}$$

$$E\{(x(n-m_i)-1)x^3(n-m_i)\}=3 \tag{8A.2.7.d}$$

Thus, $\mathbf{R}_{\tilde{X}X}$ can be expressed explicitly as:

$$\mathbf{R}_{\tilde{X}X} = \begin{bmatrix} \mathbf{I}_{M\times M} & & & \underbrace{3\mathbf{I}_{M\times M}}_{(A.2.7.d)} & \mathbf{A}_{M\times M(M-1)} & \\ & \underbrace{2\mathbf{I}_{M\times M}}_{(A.2.7.a)} & & & & \\ & & \mathbf{I}_{\frac{M(M-1)}{2}\times\frac{M(M-1)}{2}} & & \mathbf{0} & \\ & & & \underbrace{6\mathbf{I}_{M\times M}}_{(A2.7.b)} & & \\ & \mathbf{0} & & & \underbrace{2\mathbf{I}_{M(M-1)\times M(M-1)}}_{(A.2.7.c)} & \\ & & & & & \mathbf{I}_{C(M,3)\times C(M,3)} \end{bmatrix}_{L\times L} \qquad (8A.2.8)$$

where $\mathbf{I}_{p\times q}$ is a $p\times q$ identity matrix. According to equation 8.1 and equation 8.4, $\mathbf{A}_{M\times M(M-1)}$ is calculated by:

$$\mathbf{A}_{M\times M(M-1)} = E\{\begin{bmatrix} x(n) \\ x(n-1) \\ \vdots \\ x(n-M+1) \end{bmatrix} [x(n)x^2(n-1), x^2(n)x(n-1), x(n)x^2(n-2), \ldots, x^2(n-M+2)x(n-M+1)]\}$$

$$= \begin{bmatrix} 1 & 0 & 1 & \ldots & 0 & 0 \\ 0 & 1 & 0 & \ldots & 0 & 0 \\ & & & \ldots & & \\ 0 & 0 & 0 & \ldots & 0 & 1 \end{bmatrix}_{M\times M(M-1)} \qquad (8A.2.9)$$

Example: If M=3, $\mathbf{A}_{3\times 6}$ has two 1's in each row which is

$$\mathbf{A}_{3\times 6} = \begin{bmatrix} 1 & 0 & 1 & 0 & 0 & 0 \\ 0 & 1 & 0 & 0 & 1 & 0 \\ 0 & 0 & 0 & 1 & 0 & 1 \end{bmatrix} \qquad (8A.2.10)$$

Substituting equation 8A.2.4, equation 8A.2.6, and equation 8A.2.8 on the right hand side of equation 8A.2.1 yields the following result:

$$\mathbf{R}_{\tilde{\mathbf{X}}\mathbf{X}}^{T}\,\mathbf{R}_{\tilde{\mathbf{X}}\tilde{\mathbf{X}}}^{-T}\,\mathbf{R}_{\tilde{\mathbf{X}}\mathbf{X}} + \mathbf{M}_x\mathbf{M}_x^{T}$$

$$= \begin{bmatrix} \mathbf{I}_{M\times M} & & & 3\mathbf{I}_{M\times M} & \mathbf{A}_{M\times M(M-1)} & \\ & \mathbf{1}_{M\times M}+2\mathbf{I}_{M\times M} & & & & \\ & & \mathbf{I}_{\frac{M(M-1)}{2}\times\frac{M(M-1)}{2}} & & \mathbf{0} & \\ & & & 15\mathbf{I}_{M\times M} & 3\mathbf{A}_{M\times M(M-1)} & \\ & \text{symmetric} & & & \mathbf{A}'_{M(M-1)\times M(M-1)} & \\ & & & & & \mathbf{I}_{C(M,3)\times C(M,3)} \end{bmatrix}_{L\times L} \tag{8A.2.11}$$

where

$$\mathbf{A}'_{M(M-1)\times M(M-1)} = \mathbf{A}^{T}_{M\times M(M-1)}\mathbf{A}_{M\times M(M-1)} + 2\mathbf{I}_{M(M-1)\times M(M-1)}$$
$$= \begin{bmatrix} 3 & 0 & 1 & \dots & 0 & 0 \\ 0 & 3 & 0 & \dots & 0 & 0 \\ & & & \dots & & \\ 0 & 0 & 0 & \dots 1 \dots & 0 & 3 \end{bmatrix}_{M(M-1)\times M(M-1)} \tag{8A.2.12}$$

Note that in equation 8A.2.12 there are M–1 non-zero terms in each row which are one 3's and (M–2) 1's. As in the previous step, we need to express matrix $\mathbf{R}_{XX}$ explicitly, which is:

$$\mathbf{R}_{XX} = \begin{bmatrix} \mathbf{I}_{M\times M} & & & 3\mathbf{I}_{M\times M} & \mathbf{A}_{M(M-1)\times M} & \\ & \mathbf{B}_{M\times M} & & \mathbf{0} & & \\ & & \mathbf{I}_{\frac{M(M-1)}{2}\times\frac{M(M-1)}{2}} & & \dots & \\ & & & 15\mathbf{I}_{M\times M} & \mathbf{D}_{M\times M(M-1)} & \\ & \text{symmetric} & & & \mathbf{C}_{M(M-1)\times M(M-1)} & \mathbf{0} \\ & & & & & \mathbf{I}_{C(M,3)\times C(M,3)} \end{bmatrix}_{L\times L} \tag{8A.2.13}$$

Note that $E\{x^6(n-m_i)\}=15$ and $E\{x^4(n-m_i)\}=3$. The sub-matrix $\mathbf{B}_{M\times M}$ in equation 8A.2.13 is calculated by:

$$\mathbf{B}_{M\times M} = E\{\begin{bmatrix} x^2(n) \\ x^2(n-1) \\ \vdots \\ x^2(n-M+1) \end{bmatrix} [x^2(n), x^2(n-1), ..., x^2(n-M+1)]\}$$

$$= \begin{bmatrix} 3 & 1 & \dots & 1 \\ 1 & 3 & & \\ & & \ddots & \\ 1 & 1 & \dots & 3 \end{bmatrix}_{M\times M}$$

$$= \mathbf{1}_{M\times M} + 2\mathbf{I}_{M\times M} \qquad (8A.2.14)$$

The sub-matrix $\mathbf{C}_{M(M-1)\times M(M-1)}$ in equation(8A.2.13) is calculated by:

$$\mathbf{C}_{M(M-1)\times M(M-1)} = E\{\begin{bmatrix} x(n)x^2(n-1) \\ x^2(n)x(n-1) \\ \vdots \\ x^2(n-M+2)x(n-M+1) \end{bmatrix} [x(n)x^2(n-1), x^2(n)x(n-1),$$

$$..., x^2(n-M+2)x(n-M+1)]\}$$

$$= \begin{bmatrix} 3 & 0 & 1 & \dots & 0 & 0 \\ 0 & 3 & 0 & \dots & 0 & 0 \\ & & & \dots & & \\ 0 & 0 & 0 & \dots 1 \dots & 0 & 3 \end{bmatrix}_{M(M-1)\times M(M-1)} \qquad (8A.2.15)$$

Note that in $\mathbf{C}_{M(M-1)\times M(M-1)}$ there are (M–1) non-zero terms in each row that includes one of $E\{x^4(n-m_i)x^2(n-m_j)\}_{i\neq j}=3$ and (M–2) of $E\{x^2(n-m_i)x^2(n-m_j)x^2(n-m_k)\}_{i\neq j\neq k}=1$. Observe that equation 8A.2.12 and equation 8A.2.15 are exactly the same. Therefore,

$$\mathbf{C}_{M(M-1)\times M(M-1)} = \mathbf{A}'_{M(M-1)\times M(M-1)} \qquad (8A.2.16)$$

Example: If M=3, $\mathbf{C}_{6\times6}$ can be expressed in terms of $\mathbf{A}_{3\times6}$

$$\mathbf{C}_{6\times6}=\begin{bmatrix}3&0&1&0&0&0\\0&3&0&0&1&0\\1&0&3&0&0&0\\0&0&0&3&0&1\\0&1&0&0&3&0\\0&0&0&1&0&3\end{bmatrix}=\begin{bmatrix}1&0&0\\0&1&0\\1&0&0\\0&0&1\\0&1&0\\0&0&1\end{bmatrix}\begin{bmatrix}1&0&1&0&0&0\\0&1&0&0&1&0\\0&0&0&1&0&1\end{bmatrix}+2\mathbf{I}_{6\times6}$$

$$\mathbf{I}_{6\times6}=\mathbf{A}'_{6\times6} \tag{8A.2.17}$$

The sub-martix

$$\mathbf{D}_{M\times(M-1)}=E\{\begin{bmatrix}x^3(n)\\x^3(n-1)\\\vdots\\x^3(n-M+1)\end{bmatrix}[x(n)x^2(n-1),\ x(n)x^2(n-1),\ x(n)x^2(n-2),\ \ldots,\ x^2(n-M+2)x(n-M+1)]\}$$

$$=\begin{bmatrix}3&0&3&\ldots&0&0\\0&3&0&\ldots&0&0\\&&&\ldots&&\\0&0&0&\ldots1\ldots&0&3\end{bmatrix}_{M\times M(M-1)}=3\,\mathbf{A}_{M\times M(M-1)} \tag{8A.2.18}$$

Note that there are (M–1) 3's in each row of $\mathbf{D}_{M\times M(M-1)}$ because $E\{x^4(n-m_i)x^2(n-m_j)\}_{i\neq j}=3$. Substituting equation 8A.2.14, equation 8A.2.16, and equation 8A.2.18 in equation 8A.2.13 yields the following equation:

$$\mathbf{R}_{xx} = \begin{bmatrix} \mathbf{I}_{M\times M} & & & \mathbf{3I}_{M\times M} & \mathbf{A}_{M\times M(M-1)} & \\ & \mathbf{1}_{M\times M}+\mathbf{2I}_{M\times M} & & & & \\ & & \mathbf{I}_{\frac{M(M-1)}{2}\times\frac{M(M-1)}{2}} & & \mathbf{0} & \\ & & & 15\mathbf{I}_{M\times M} & \mathbf{3A}_{M\times M(M-1)} & \\ & \text{symmetric} & & & \mathbf{A}'_{M(M-1)\times M(M-1)} & \\ & & & & & \mathbf{I}_{C(M,3)\times C(M,3)} \end{bmatrix}_{L\times L} \tag{8A.2.19}$$

Compare equation 8A.2.11 and equation 8A.2.19. Notice that they are indeed identical. To illustrate the numerical example of equation 8A.2.1, assume M=2; therefore, we have 9 elements in each $\mathbf{X}(n)$ and $\widetilde{\mathbf{X}}(n)$ which are:

$$\mathbf{X}(n) = [x(n), x(n-1), x^2(n), x^2(n-1), x(n)x(n-1), x^3(n), x^3(n-1), x(n)x^2(n-1), x^2(n)x(n-1)]^T$$

$$\widetilde{\mathbf{X}}(n) = [x(n), x(n-1), (x^2(n)-1), (x^2(n-1)-1), x(n)x(n-1), x^3(n)-3x(n), x^3(n-1)-3x(n-1), x(n)(x^2(n-1)-1), (x^2(n)-1)x(n-1)]^T$$

Thus,

$$\mathbf{R}_{XX} = \begin{bmatrix} 1 & 0 & 0 & 0 & 0 & 3 & 0 & 1 & 0 \\ 0 & 1 & 0 & 0 & 0 & 0 & 3 & 0 & 1 \\ 0 & 0 & 3 & 1 & 0 & 0 & 0 & 0 & 0 \\ 0 & 0 & 1 & 3 & 0 & 0 & 0 & 0 & 0 \\ 0 & 0 & 0 & 0 & 1 & 0 & 0 & 0 & 0 \\ 3 & 0 & 0 & 0 & 0 & 15 & 0 & 3 & 0 \\ 0 & 3 & 0 & 0 & 0 & 0 & 15 & 0 & 3 \\ 1 & 0 & 0 & 0 & 0 & 3 & 0 & 3 & 0 \\ 0 & 1 & 0 & 0 & 0 & 0 & 3 & 0 & 3 \end{bmatrix},$$

$$\mathbf{R}_{\tilde{\mathbf{X}}\mathbf{X}} = \begin{bmatrix} 1 & 0 & 0 & 0 & 0 & 3 & 0 & 1 & 0 \\ 0 & 1 & 0 & 0 & 0 & 0 & 3 & 0 & 1 \\ 0 & 0 & 2 & 0 & 0 & 0 & 0 & 0 & 0 \\ 0 & 0 & 0 & 2 & 0 & 0 & 0 & 0 & 0 \\ 0 & 0 & 0 & 0 & 1 & 0 & 0 & 0 & 0 \\ 0 & 0 & 0 & 0 & 0 & 6 & 0 & 0 & 0 \\ 0 & 0 & 0 & 0 & 0 & 0 & 6 & 0 & 0 \\ 0 & 0 & 0 & 0 & 0 & 0 & 0 & 2 & 0 \\ 0 & 0 & 0 & 0 & 0 & 0 & 0 & 0 & 2 \end{bmatrix}$$

$$\mathbf{R}_{\tilde{\mathbf{X}}\tilde{\mathbf{X}}} = \begin{bmatrix} 1 & 0 & 0 & 0 & 0 & 0 & 0 & 0 & 0 \\ 0 & 1 & 0 & 0 & 0 & 0 & 0 & 0 & 0 \\ 0 & 0 & 2 & 0 & 0 & 0 & 0 & 0 & 0 \\ 0 & 0 & 0 & 2 & 0 & 0 & 0 & 0 & 0 \\ 0 & 0 & 0 & 0 & 1 & 0 & 0 & 0 & 0 \\ 0 & 0 & 0 & 0 & 0 & 6 & 0 & 0 & 0 \\ 0 & 0 & 0 & 0 & 0 & 0 & 6 & 0 & 0 \\ 0 & 0 & 0 & 0 & 0 & 0 & 0 & 2 & 0 \\ 0 & 0 & 0 & 0 & 0 & 0 & 0 & 0 & 2 \end{bmatrix},$$

$$\mathbf{M}_{\mathbf{X}}\mathbf{M}_{\mathbf{X}}^{\mathbf{T}} = \begin{bmatrix} 0 & 0 & 0 & 0 & 0 & 0 & 0 & 0 & 0 \\ 0 & 0 & 0 & 0 & 0 & 0 & 0 & 0 & 0 \\ 0 & 0 & 1 & 1 & 0 & 0 & 0 & 0 & 0 \\ 0 & 0 & 1 & 1 & 0 & 0 & 0 & 0 & 0 \\ 0 & 0 & 0 & 0 & 0 & 0 & 0 & 0 & 0 \\ 0 & 0 & 0 & 0 & 0 & 0 & 0 & 0 & 0 \\ 0 & 0 & 0 & 0 & 0 & 0 & 0 & 0 & 0 \\ 0 & 0 & 0 & 0 & 0 & 0 & 0 & 0 & 0 \\ 0 & 0 & 0 & 0 & 0 & 0 & 0 & 0 & 0 \end{bmatrix}$$

Substitute the above matrices in equation 8A.2.1 and the correctness is clearly demonstrated.

Comparing the third-order case with the second-order case, we find that if $\mathbf{R}_{\tilde{\mathbf{X}}\mathbf{X}} = \mathbf{R}_{\tilde{\mathbf{X}}\tilde{\mathbf{X}}}$ or $\mathbf{R}_{\tilde{\mathbf{X}}\mathbf{X}}\mathbf{R}_{\tilde{\mathbf{X}}\tilde{\mathbf{X}}}^{-1} = \mathbf{I}$, then equation 8A.2.1 becomes:

$$\mathbf{R}_{\mathbf{XX}} = \mathbf{R}_{\tilde{\mathbf{X}}\mathbf{X}}^{\mathrm{T}}\,\mathbf{R}_{\tilde{\mathbf{X}}\tilde{\mathbf{X}}}^{-\mathrm{T}}\,\mathbf{R}_{\tilde{\mathbf{X}}\mathbf{X}} + \mathbf{M}_x\mathbf{M}_x^{\mathrm{T}} = \mathbf{R}_{\tilde{\mathbf{X}}\mathbf{X}} + \mathbf{M}_x\mathbf{M}_x^{\mathrm{T}} \qquad (8A.2.20)$$

This means that equation 8A.2.1 is the higher-order extension version of equation 7A.1.14. In other words, we can say that equation 7A.1.14 is the special case of equation 8A.2.1, through the proper manipulations of equation 8A.2.20 as follows:

$$\mathbf{R}_{\tilde{X}X} = (\mathbf{R}_{\tilde{X}X}^{T}\,\mathbf{R}_{\tilde{X}\tilde{X}}^{-T})^{-1}\,(\mathbf{R}_{XX} - \mathbf{M}_x\mathbf{M}_x^{T})$$

$$\mathbf{R}_{\tilde{X}X}\,\mathbf{R}_{\tilde{X}X}^{-1}\,\mathbf{R}_{\tilde{X}\tilde{X}} = (\mathbf{R}_{\tilde{X}X}^{-1}\,\mathbf{R}_{\tilde{X}\tilde{X}})^{T}\,(\mathbf{R}_{XX} - \mathbf{M}_x\mathbf{M}_x^{T})\,\mathbf{R}_{\tilde{X}X}^{-1}\,\mathbf{R}_{\tilde{X}\tilde{X}}$$

Therefore, there is an alternative way to express the relation of $\mathbf{R}_{XX}$, $\mathbf{R}_{\tilde{X}X}$ and $\mathbf{R}_{\tilde{X}\tilde{X}}$, which is:

$$\mathbf{R}_{\tilde{X}\tilde{X}} = (\mathbf{R}_{\tilde{X}X}^{-1}\,\mathbf{R}_{\tilde{X}\tilde{X}})^{T}\,(\mathbf{R}_{XX} - \mathbf{M}_x\mathbf{M}_x^{T})\,\mathbf{R}_{\tilde{X}X}^{-1}\,\mathbf{R}_{\tilde{X}\tilde{X}} \qquad (8A.2.21)$$

Equation 8A.2.21 is used to prove the misadjustment property of third-order nonlinear Wiener LMS adaptive filter in chapter 8.

8.5 Appendix 8B: Inverse Matrix of the Cross-Correlation Matrix $\mathbf{R}_{\tilde{X}X}$

Consider a matrix lower-triangular matrix **R** which can be partitioned into several sub-matrices described as follows:

$$\mathbf{R} = \begin{bmatrix} \mathbf{D}_{11} & \mathbf{R}_{13} & \mathbf{D}_{13} \\ \mathbf{0} & \mathbf{D}_{22} & \mathbf{0} \\ \mathbf{0} & \mathbf{0} & \mathbf{I}_{33} \end{bmatrix} \qquad (8B.3.1)$$

where $\mathbf{D}_{11}$, $\mathbf{D}_{22}$, and $\mathbf{D}_{31}$ are diagonal matrices, and $\mathbf{I}_{33}$ is an identity matrix. According to the inversion lemma one, the inverse matrix of **R** can be found as:

$$\mathbf{R}^{-1} = \begin{bmatrix} \mathbf{D}_{11}^{-1} & \mathbf{R}'_{12} & \mathbf{D}'_{13} \\ \mathbf{0} & \mathbf{D}_{22}^{-1} & \mathbf{0} \\ \mathbf{0} & \mathbf{0} & \mathbf{I}_{33} \end{bmatrix} \qquad (8B.3.2)$$

where the sub-matrix $\begin{bmatrix} \mathbf{R}'_{12} & \mathbf{D}'_{13} \end{bmatrix}$ is:

$$\begin{bmatrix}\mathbf{R}'_{12} & \mathbf{D}'_{13}\end{bmatrix} = -\begin{bmatrix}\mathbf{R}_{12} & \mathbf{D}_{13}\end{bmatrix}\begin{bmatrix}\mathbf{D}_{11}^{-1} & \mathbf{0}\\ \mathbf{0} & \mathbf{D}_{22}^{-1}\end{bmatrix} = -\begin{bmatrix}\mathbf{R}_{12}\mathbf{D}_{11}^{-1} & \mathbf{D}_{13}\mathbf{D}_{22}^{-1}\end{bmatrix} \qquad (8B.3.3)$$

Observe equation 8B.3.2 and equation 8B.3.3. Note that $\mathbf{R}^{-1}$ can be easily found simply by inspection. This means that all the diagonal elements in $\mathbf{R}^{-1}$ are just the inverse of the original number, and all the off-diagonal elements in $\mathbf{R}^{-1}$ are equal to the negative value of the original number divided by the corresponding diagonal element of $\mathbf{R}$ in each particular column and the diagonal element of $\mathbf{D}_{33}$ in each particular row.

Compare equation 8A.2.9 with equation 8B.3.1. We note that $\mathbf{R}_{\tilde{X}X}$ is a special case of $\mathbf{R}$. Therefore, by using the inspection method, $\mathbf{R}_{\tilde{X}X}^{-1}$ can be easily found. For instance, the inverse of $\mathbf{R}_{\tilde{X}X}$ in the previous section can be found immediately to be:

$$\mathbf{R}_{\tilde{X}X}^{-1} = \begin{bmatrix}1 & 0 & 0 & 0 & 0 & \frac{-3}{6} & 0 & \frac{-1}{2} & 0\\ 0 & 1 & 0 & 0 & 0 & 0 & \frac{-3}{6} & 0 & \frac{-1}{2}\\ 0 & 0 & \frac{1}{2} & 0 & 0 & 0 & 0 & 0 & 0\\ 0 & 0 & 0 & \frac{1}{2} & 0 & 0 & 0 & 0 & 0\\ 0 & 0 & 0 & 0 & \frac{1}{2} & 0 & 0 & 0 & 0\\ 0 & 0 & 0 & 0 & 0 & \frac{1}{6} & 0 & 0 & 0\\ 0 & 0 & 0 & 0 & 0 & 0 & \frac{1}{6} & 0 & 0\\ 0 & 0 & 0 & 0 & 0 & 0 & 0 & \frac{1}{2} & 0\\ 0 & 0 & 0 & 0 & 0 & 0 & 0 & 0 & \frac{1}{2}\end{bmatrix}$$

8.6 Appendix 8C: Verification of Equation 8.16

To demonstrate that $\xi(n)$ can be expressed as a quadratic form of $\mathbf{V}(n)$

$$\xi(n) = \xi_{min} + \mathbf{V}^{T}(n)\mathbf{R}_{\tilde{Q}\tilde{Q}}\mathbf{V}(n) \qquad (8C.1)$$

first, substitute equation 8.14 in the second term of equation 8C.1 and reduce to a total of 9 terms, which are:

$$\mathbf{S}_{\tilde{Q}}^{-1}\mathbf{C}^{T}(n)\begin{bmatrix}1 & \mathbf{0}^{T}\\ \mathbf{0} & \mathbf{R}_{\tilde{X}\tilde{X}}\end{bmatrix}\mathbf{S}_{\tilde{Q}}^{-1}\mathbf{C}(n) = w_0^2(n) + \mathbf{W}^{T}(n)\mathbf{S}_{\tilde{X}}^{-1}\mathbf{R}_{\tilde{X}\tilde{X}}\mathbf{S}_{\tilde{X}}^{-1}\mathbf{W}(n) \qquad (8C.2)$$

$$\mathbf{C}^{*T}\begin{bmatrix}1 & \mathbf{0}^T\\ \mathbf{0} & \mathbf{R}_{\tilde{X}\tilde{X}}^{-1}\mathbf{R}_{\tilde{X}X}\end{bmatrix}\begin{bmatrix}1 & \mathbf{0}^T\\ \mathbf{0} & \mathbf{R}_{\tilde{X}\tilde{X}}\end{bmatrix}\mathbf{S}_{\tilde{Q}}^{-1}\mathbf{C}(n) = -w_0^{*} w_0(n) - \mathbf{W}^T(n)\mathbf{S}_{\tilde{X}}^{-1}\mathbf{R}_{\tilde{X}X}\mathbf{W}^{*} \tag{8C.3}$$

$$-\begin{bmatrix}E\{\mathbf{X}^T(n)\}\mathbf{W}^{*}\\ \mathbf{0}\end{bmatrix}^T\begin{bmatrix}1 & \mathbf{0}^T\\ \mathbf{0} & \mathbf{R}_{\tilde{X}\tilde{X}}\end{bmatrix}\mathbf{S}_{\tilde{Q}}^{-1}\mathbf{C}(n) = -w_0(n)E\{\mathbf{X}^T(n)\}\mathbf{W}^{*} \tag{8C.4}$$

$$-(\mathbf{S}_{\tilde{Q}}^{-1}\mathbf{C}^T(n))^T\begin{bmatrix}1 & \mathbf{0}^T\\ \mathbf{0} & \mathbf{R}_{\tilde{X}\tilde{X}}\end{bmatrix}\begin{bmatrix}1 & \mathbf{0}^T\\ \mathbf{0} & \mathbf{R}_{\tilde{X}\tilde{X}}^{-1}\mathbf{R}_{\tilde{X}X}\end{bmatrix}\mathbf{C}^{*} = -w_0(n)\, w_0^{*} - \mathbf{W}^T(n)\mathbf{S}_{\tilde{X}}^{-1}\mathbf{R}_{\tilde{X}X}\mathbf{W}^{*} \tag{8C.5}$$

$$\left(\begin{bmatrix}1 & \mathbf{0}^T\\ \mathbf{0} & \mathbf{R}_{\tilde{X}\tilde{X}}^{-1}\mathbf{R}_{\tilde{X}X}\end{bmatrix}\mathbf{C}^{*}\right)^T\begin{bmatrix}1 & \mathbf{0}^T\\ \mathbf{0} & \mathbf{R}_{\tilde{X}\tilde{X}}\end{bmatrix}\begin{bmatrix}1 & \mathbf{0}^T\\ \mathbf{0} & \mathbf{R}_{\tilde{X}\tilde{X}}^{-1}\mathbf{R}_{\tilde{X}X}\end{bmatrix}\mathbf{C}^{*}$$

$$= [\,w_0^{*}\ \mathbf{W}^{*T}]\begin{bmatrix}1 & \mathbf{0}^T\\ \mathbf{0} & \mathbf{R}_{\tilde{X}\tilde{X}}^{-1}\mathbf{R}_{\tilde{X}X}\end{bmatrix}^T\begin{bmatrix}w_0^{*}\\ \mathbf{R}_{\tilde{X}X}\mathbf{W}^{*}\end{bmatrix}$$

$$= w_0^{*2} + \mathbf{W}^{*T}\mathbf{R}_{\tilde{X}X}^T\mathbf{R}_{\tilde{X}\tilde{X}}^{-T}\mathbf{R}_{\tilde{X}X}\mathbf{W}^{*}$$

$$= w_0^{*2} + \mathbf{W}^{*T}(\mathbf{R}_{xx} - \mathbf{M}_x\mathbf{M}_x^T)\mathbf{W}^{*} = w_0^{*2} + \mathbf{W}^{*T}\mathbf{R}_{xx}\mathbf{W}^{*} - \mathbf{W}^{*T}\mathbf{M}_x\mathbf{M}_x^T\mathbf{W}^{*} \tag{8C.6}$$

$$\begin{bmatrix}E\{\mathbf{X}^T(n)\}\mathbf{W}^{*}\\ \mathbf{0}\end{bmatrix}^T\begin{bmatrix}1 & \mathbf{0}^T\\ \mathbf{0} & \mathbf{R}_{\tilde{X}\tilde{X}}\end{bmatrix}\begin{bmatrix}1 & \mathbf{0}^T\\ \mathbf{0} & \mathbf{R}_{\tilde{X}\tilde{X}}^{-1}\mathbf{R}_{\tilde{X}X}\end{bmatrix}\mathbf{C}^{*} = -w_0^{*}E\{\mathbf{X}^T(n)\}\mathbf{W}^{*} \tag{8C.7}$$

$$-(\mathbf{S}_{\tilde{Q}}^{-1}\mathbf{C}^T(n))^T\begin{bmatrix}1 & \mathbf{0}^T\\ \mathbf{0} & \mathbf{R}_{\tilde{X}\tilde{X}}\end{bmatrix}\begin{bmatrix}E\{\mathbf{X}^T(n)\}\mathbf{W}^{*}\\ \mathbf{0}\end{bmatrix}\mathbf{C}^{*} = -w_0^{*}E\{\mathbf{X}^T(n)\}\mathbf{W}^{*} \tag{8C.8}$$

$$\left(\begin{bmatrix}1 & \mathbf{0}^T\\ \mathbf{0} & \mathbf{R}_{\tilde{X}\tilde{X}}^{-1}\mathbf{R}_{\tilde{X}X}\end{bmatrix}\mathbf{C}^{*}\right)^T\begin{bmatrix}1 & \mathbf{0}^T\\ \mathbf{0} & \mathbf{R}_{\tilde{X}\tilde{X}}\end{bmatrix}\begin{bmatrix}E\{\mathbf{X}^T(n)\}\mathbf{W}^{*}\\ \mathbf{0}\end{bmatrix} = w_0^{*}E\{\mathbf{X}^T(n)\}\mathbf{W}^{*} \tag{8C.9}$$

$$\begin{bmatrix}E\{\mathbf{X}^T(n)\}\mathbf{W}^{*}\\ \mathbf{0}\end{bmatrix}^T\begin{bmatrix}1 & \mathbf{0}^T\\ \mathbf{0} & \mathbf{R}_{\tilde{X}\tilde{X}}\end{bmatrix}\begin{bmatrix}E\{\mathbf{X}^T(n)\}\mathbf{W}^{*}\\ \mathbf{0}\end{bmatrix} = (E\{\mathbf{X}^T(n)\}\mathbf{W}^{*})^2 \tag{8C.10}$$

Note that the expression of equation 8C.6 is expanded according to equation 8A.2.1. The second step replaces the quadratic term in equation 8C.1 by the summation of equation 8C.2- 8C.10. In steady state, we have **C** = [w_0 $\mathbf{W}^T$]^T, which yields:

$$\begin{aligned}
\xi &= \xi_{min} + (w_0^{*2} - 2w_0 w_0^* + w_0^2) + 2w_0^* E\{\mathbf{X}^T(n)\}\mathbf{W}^* - 2w_0 E\{\mathbf{X}^T(n)\}\mathbf{W}^* \\
&\qquad + \mathbf{W}^{*T}\mathbf{R}_{XX}\mathbf{W}^* \\
&\quad + \mathbf{W}^T\mathbf{S}_{\tilde{X}}^{-1}\,\mathbf{R}_{\tilde{X}\tilde{X}}\mathbf{S}_{\tilde{X}}^{-1}\,\mathbf{W}(n) - 2\mathbf{W}^T\mathbf{S}_{\tilde{X}}^{-1}\mathbf{R}_{\tilde{X}X}\mathbf{W}^* \\
&= \xi_{min} + (w_0^* - w_0)^2 + 2w_0^* E\{\mathbf{X}^T(n)\}\mathbf{W}^* - 2w_0 E\{\mathbf{X}^T(n)\}\mathbf{W}^* + \mathbf{W}^{*T}\mathbf{R}_{XX}\mathbf{W}^* \\
&\quad + \mathbf{W}^T\mathbf{S}_{\tilde{X}}^{-1}\,\mathbf{R}_{\tilde{X}\tilde{X}}\mathbf{S}_{\tilde{X}}^{-1}\,\mathbf{W} - 2\mathbf{W}^T\mathbf{S}_{\tilde{X}}^{-1}\mathbf{R}_{\tilde{X}X}\mathbf{W}^* \qquad (8C.11)
\end{aligned}$$

The result of equation 8C.11 corresponds to equation 8.9 and thus validates equation 8.16.

Chapter 9

NONLINEAR ADAPTIVE SYSTEM IDENTIFICATION BASED ON WIENER MODELS (PART 3)

Other stochastic-gradient-based algorithms

Introduction

In this third part of the book focused on nonlinear adaptive system identification algorithms based on the Wiener model, we discuss some algorithms which are suitable for situations where the environment leads to a non-white, possibly non-Gaussian input signal. We also discuss using other stochastic-gradient-based algorithms like the least-mean-fourth (LMF) algorithm for the Wiener model.

9.1 Nonlinear LMF Adaptation Algorithm

From the detailed analysis in the previous two chapters, we note that the nonlinear Wiener LMS adaptive filter is a direct extension of the linear LMS algorithm. The advantage is that most of the linear properties are inherited, even though we are dealing with a complicated nonlinear system. Therefore, based on the background in these previous chapters, further extension to nonlinear LMF algorithm is possible, as we now show.

Compared to the LMS method, which minimizes the expected value of the square of the error, $E\{e^2(n)\}$, it is more general to minimize the $E\{e^{2K}(n)\}$ criterion as presented in Walach (1984). This allows us to minimize the measurement error in the mean fourth and mean sixth etc., sense. The general form of the algorithms is:

$$\mathbf{C}(n+1) = \mathbf{C}(n) + 2\mu K e^{2K-1}(n)\,\tilde{\mathbf{Q}}(n) \qquad (9.1a)$$

As for the LMS,, the range of μ, the time constant, and misadjustment are:

$$0<\mu<\frac{1}{KL(2K-1)E\{n^{2K-2}(n)\}E\{x^2(n)\}} \tag{9.1b}$$

$$\tau_i=\frac{1}{2\mu K(2K-1)E\{n^{2K-2}(n)\}\lambda_i} \tag{9.2}$$

and

$$MISADJ=\frac{E[n^{4K-2}(n)]}{2(2K-1)^2E[n^2(n)]\left(E[n^{2K-2}(n)]\right)^2}\sum_{i=1}^{L}\frac{1}{\tau_i} \tag{9.3}$$

To compare the performance, define r(K) which is the ratio of equation 7.40 and equation 9.3:

$$r(K)=\frac{(2K-1)^2E[n^2(n)]E[n^{2K-2}(n)]}{E[n^{4K-2}(n)]} \tag{9.4}$$

When r(K) > 1, the algorithm of K >1 will be advantageous over that of LMS if the plant noise is non-Gaussian. This means that lower misadjustment can be expected for the same speed of convergence.

9.2 Transform Domain Nonlinear Wiener Adaptive Filter

Although all the results shown in the two previous chapters assume that the input signal is Gaussian white noise, it is worth noting that this algorithm can be extended to any independent and identically distributed non-Gaussian signal.

For the colored input signal, in continuous-time domain, a whitening function may be used to generate L-functional (Schetzen 1980) to decorrelate the input signal. For the discrete-time domain, the whitening filter can be implemented by a lattice structure (Haykin 1996) if the number of stages is long enough. Another way to achieve the whitening effect is to use the TD (transform domain) algorithm (Narayan 1983) to decouple the correlated colored data. The original block diagram in chapter 7 can be modified as in figure 9-1.

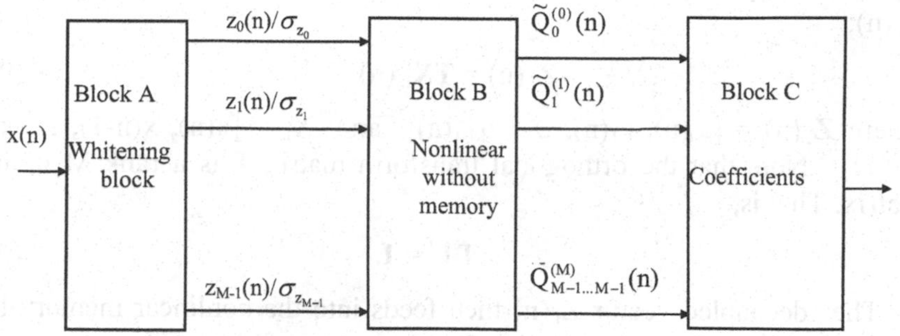

Figure 9-1. Nonlinear discrete Wiener model for colored input

Note that, in figure 9-1, if x(n) can be perfectly decorrelated by a whitening filter, it implies that each of the normalized $z_i(n)$ is mutually independent and identically distributed (i.i.d.) to each other. Therefore, the same nonlinear LMS algorithm can be applied to identify the Volterra system with colored input. The effectiveness of this whitening will be clearly demonstrated by computer simulations later.

To investigate the condition number improvement, consider the second-order TD nonlinear discrete Wiener system shown in figure 9-2 (Chang 2000a):

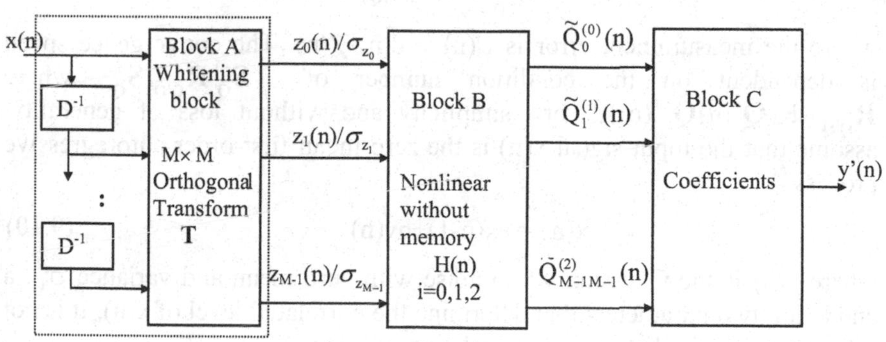

Figure 9-2 Transform domain second-order nonlinear discrete Wiener system

The colored input vector $\mathbf{X}_L(n)$ is first transformed into another vector $\mathbf{Z}_L(n)$:

$$\mathbf{Z}_L(n) = \mathbf{T}\mathbf{X}_L(n) \tag{9.5}$$

where $\mathbf{Z}_L(n) = [z_0(n), z_1(n), ..., z_{M-1}(n)]^T$ and $\mathbf{X}_L = [x(n), x(n-1), ..., x(n-M+1)]^T$. Note that the orthogonal transform matrix $\mathbf{T}$ is a rank M unitary matrix. That is,

$$\mathbf{T}\mathbf{T}^T = \mathbf{I}. \tag{9.6}$$

The decoupled vector $\mathbf{Z}_L(n)$ then feeds into the nonlinear memory-less processing block B. As in equation 7.9, the output of block C becomes:

$$y'(n) = \mathbf{C}'^T(n)\, \mathbf{S}_{\tilde{\mathbf{Q}}}^{-1}\, \tilde{\mathbf{Q}}'(n) = \mathbf{C}'^T(n)\, \mathbf{S}_{\tilde{\mathbf{Q}}}^{-1}\, [1, \tilde{\mathbf{Z}}^T(n)]^T = w_0' + \mathbf{W}'^T(n)\, \mathbf{S}_{\tilde{\mathbf{Z}}}^{-1}\, \tilde{\mathbf{Z}}(n) \tag{9.7}$$

The vector $\tilde{\mathbf{Z}}(n)$ is defined as:

$$\tilde{\mathbf{Z}}(n) = [z_0(n), z_1(n), ..., z_{M-1}(n), z_1^2(n)-\sigma_{z_1}^2, ..., z_{M-1}^2(n)-\sigma_{z_{M-1}}^2, z_0(n)z_1(n), ..., z_{M-2}(n)z_{M-1}(n)]^T \tag{9.8}$$

Assume that the $M\times M$ transform matrix can perfectly decouple the Gaussian colored input to an i.i.d. signal. If the power normalization of each output bin of block A can be done properly then, followed by equation 7.65, the optimal weight vector can be obtained by :

$$\mathbf{C}'(n+1) = \mathbf{C}'(n) + 2\mu e'(n)\, \mathbf{S}_{\tilde{\mathbf{Q}}'}^{-1}\, \tilde{\mathbf{Q}}'(n) \tag{9.9}$$

where the measurement error is $e'(n) = d(n)-y'(n)$. The convergence speed is dependent on the condition number of $\mathbf{S}_{\tilde{\mathbf{Q}}'}^{-1}\mathbf{R}_{\tilde{\mathbf{Q}}'\tilde{\mathbf{Q}}'}\mathbf{S}_{\tilde{\mathbf{Q}}'}^{-1}$, where $\mathbf{R}_{\tilde{\mathbf{Q}}'\tilde{\mathbf{Q}}'}=E\{\tilde{\mathbf{Q}}'(n)\tilde{\mathbf{Q}}'^T(n)\}$. For simplicity and without loss of generality, assume that the input signal $x(n)$ is the zero mean first-order autoregressive process

$$x(n) = ax(n-1)+bv(n) \tag{9.10}$$

where $v(n)$ is the Gaussian white noise with zero mean and variance σ_v^2. a and b are two parameters that determine the correlated level of $x(n)$. It is not difficult to show that the autocorrelation of x is :

$$\mathbf{R}_{xx}(k)= \frac{\sigma_v^2 b^2}{1-|a|^2}\, a^k u(k) \tag{9.11}$$

where $u(k)$ is the unit step function. And from the relationship

$$E\{x(n)x(n-1)x(n-2)x(n-3)\} = E\{x(n)x(n-1)\}E\{x(n-2)x(n-3)\} + E\{x(n)x(n-2)\}E\{x(n-1)x(n-3)\}+E\{x(n)x(n-3)\}E\{x(n-1)x(n-2)\} \tag{9.12}$$

we can express the expectation value of third- and fourth-order joint moments of x(n) as:

$$E(x^2(n)x(n-k)\} = 0 \tag{9.13a}$$

$$E\{x^2(n)x^2(n-k)\} = \frac{\sigma_v^2 b^2}{1-|a|^2} a^k + 2\left(\frac{\sigma_v^2 b^2}{1-|a|^2} a^k\right)^2 \tag{9.13b}$$

From matrix theory, the upper bound of the condition number of a square matrix **A** can be denoted as:

$$\Omega(\mathbf{A}) = \mathrm{Tr}(\mathbf{A})/\det(\mathbf{A}) \tag{9.14}$$

where Tr(.) and det(.) mean the trace and determinant. Now for first-order autoregressive process colored input with zero mean and unit power, we have:

$$\Omega(\mathbf{S}_{\tilde{Q}'}^{-1}\mathbf{R}_{\tilde{Q}'\tilde{Q}'}\mathbf{S}_{\tilde{Q}'}^{-1}) = \mathrm{Tr}(\mathbf{S}_{\tilde{Q}'}^{-1}\mathbf{R}_{\tilde{Q}'\tilde{Q}'}\mathbf{S}_{\tilde{Q}'}^{-1})/\det(\mathbf{S}_{\tilde{Q}'}^{-1}\mathbf{R}_{\tilde{Q}'\tilde{Q}'}\mathbf{S}_{\tilde{Q}'}^{-1}) = \frac{L+1}{\det(\mathbf{S}_{\tilde{Q}'}^{-1}\mathbf{R}_{\tilde{Q}'\tilde{Q}'}\mathbf{S}_{\tilde{Q}'}^{-1})} \tag{9.15}$$

This is because $\mathrm{Tr}(\mathbf{S}_{\tilde{Q}'}^{-1}\mathbf{R}_{\tilde{Q}'\tilde{Q}'}\mathbf{S}_{\tilde{Q}'}^{-1}) = L+1$, which theoretically is true if matrix **T** can really decorrelate the colored input. Without using the **T** matrix, the upper bound of $\mathbf{S}_{\tilde{Q}'}^{-1}\mathbf{R}_{\tilde{Q}'\tilde{Q}'}\mathbf{S}_{\tilde{Q}'}^{-1}$ becomes:

$$\Omega(\mathbf{S}_{\tilde{Q}}^{-1}\mathbf{R}_{\tilde{Q}\tilde{Q}}\mathbf{S}_{\tilde{Q}}^{-1}) = \frac{\mathrm{Tr}(\mathbf{S}_{\tilde{Q}}^{-1}\mathbf{R}_{\tilde{Q}\tilde{Q}}\mathbf{S}_{\tilde{Q}}^{-1})}{\det(\mathbf{S}_{\tilde{Q}}^{-1}\mathbf{R}_{\tilde{Q}\tilde{Q}}\mathbf{S}_{\tilde{Q}}^{-1})} \tag{9.16}$$

Note that, using equation 9.13a and equation 9.13b, $\mathbf{S}_{\tilde{Q}}^{-1}$ can be found which can make all the diagonal elements in $\mathbf{R}_{\tilde{Q}\tilde{Q}}$ equal to one. This means that all the diagonal elements in $\mathbf{S}_{\tilde{Q}}^{-1}\mathbf{R}_{\tilde{Q}\tilde{Q}}\mathbf{S}_{\tilde{Q}}^{-1}$ are equal to one. Therefore,

$$\mathrm{Tr}(\mathbf{S}_{\tilde{Q}}^{-1}\mathbf{R}_{\tilde{Q}\tilde{Q}}\mathbf{S}_{\tilde{Q}}^{-1}) = L+1 = \mathrm{Tr}(\mathbf{S}_{\tilde{Q}'}^{-1}\mathbf{R}_{\tilde{Q}'\tilde{Q}'}\mathbf{S}_{\tilde{Q}'}^{-1}). \tag{9.17}$$

Furthermore, $\mathbf{S}_{\tilde{Q}}^{-1}\mathbf{R}_{\tilde{Q}\tilde{Q}}\mathbf{S}_{\tilde{Q}}^{-1}$ is a positive definite matrix, which means all the off-diagonal elements are smaller than 1. This implies that

$$\det(\mathbf{S}_{\tilde{Q}}^{-1}\mathbf{R}_{\tilde{Q}\tilde{Q}}\mathbf{S}_{\tilde{Q}}^{-1}) \le \det(\mathbf{S}_{\tilde{Q}'}^{-1}\mathbf{R}_{\tilde{Q}'\tilde{Q}'}\mathbf{S}_{\tilde{Q}'}^{-1}). \tag{9.18}$$

Hence, comparing equation 9.15 and equation 9.16, we have

$$\Omega(\mathbf{S}_{\tilde{Q}'}^{-1}\mathbf{R}_{\tilde{Q}'\tilde{Q}'}\mathbf{S}_{\tilde{Q}'}^{-1}) \le \Omega(\mathbf{S}_{\tilde{Q}}^{-1}\mathbf{R}_{\tilde{Q}\tilde{Q}}\mathbf{S}_{\tilde{Q}}^{-1}) \tag{9.19}$$

From equation 9.19, we note that after properly selecting the orthonormal matrix **T**, a reduction in condition number can be achieved. For example, if the input x(n) has zero mean and unit power, the theoretical and experimental (ensemble average) upper bound values of condition number for 4-sample memory second-order nonlinear adaptive Wiener filter are listed in table 9.1:

Table 9.1. Upper bound of condition numbers of two nonlinear filters (*: Assume $a^2+b^2=1$)

	Transform domain nonlinear Wiener filter		Non-transform domain nonlinear Wiener filter	
	Theoretical	Experimental	Theoretical	Experimental
b=0.0*	15	15.340	15	15.255
b=0.2	15	16.384	39.092	40.387
b=0.5	15	17.485	1.160e4	1.192e4
b=0.9	15	19.123	2.882e16	2.361e16

From table 9.1, we know that the correlated level affects the upper bound of the condition number for TD and non-TD nonlinear adaptive Wiener filters. As b=0, there is no difference between TD and non-TD systems. When b's value is small, the upper bound of both systems increases slightly. As b becomes larger, it does not affect the upper bound value of TD systems too much, while this value grows exponentially in the non-TD system. This means that the TD system may have much better performance than the non-TD system especially in a highly colored Gaussian input environment. It can be further verified by the computer simulation in the next section.

The second-order transform domain nonlinear LMS adaptive algorithm is listed in table 9.2. A count of the arithmetic operations involved in the implementation of the algorithm shows that it requires $3M^2+5M+3$ multiplications per iteration. Our approach therefore has $O(M^2)$ computational complexity.

Table 9.2. Computational complexity of the second-order TD nonlinear Wiener LMS filter

Initialization:	$\mathbf{C}(n) = \mathbf{0}$ Precalculate $\mathbf{S}_{\tilde{\mathbf{Q}}'}^{-1}$ $\mu' = 2$		
	Relation	Dimension	Multiplication Count
Block A	$\mathbf{Z}_L(n) = \mathbf{T}\mathbf{X}_L(n)$	$M\times 1$	M^2
Block B	$\tilde{\mathbf{Q}}'(n) = [1, \tilde{\mathbf{Z}}^T(n)]^T$ $\tilde{\mathbf{Z}}(n)$ is defined in equation 7.102	$(L+1)\times 1$	$(M^2+M)/2$
Block C	$\mathbf{C}'(n+1) = \mathbf{C}'(n) + \mu' e(n) \mathbf{S}_{\tilde{\mathbf{Q}}'}^{-1}\tilde{\mathbf{Q}}'(n)$	$(L+1)\times 1$	$3(L+1)$
			Total: $3M^2+5M+3$

The count of arithmetic operations involved in the implementation of the algorithm shows that this algorithm has $O(M^2)$ computational complexity.

9.3 Computer Simulation Examples

Example 1:

Assume that an arbitrary unknown plant can be described by a 4-sample memory, second-order Volterra series. The input-output relationship (the same system as in chapter 7) is:

$$d(n) = -0.78x(n) - 1.48x(n-1) + 1.39x(n-2) + 0.04x(n-3) + 0.54x^2(n) - 1.62x^2(n-1) + 1.41x^2(n-2) - 0.13x^2(n-3) + 3.72x(n)x(n-1) + 1.86x(n)x(n-2) - 0.76x(n)x(n-3) + 0.76x(n-1)x(n-2) - 0.12x(n-1)x(n-3) - 1.52x(n-2)x(n-3) \quad (7.49)$$

The input signal x(n) is colored by the linear system with input-output relationship

$$x(n) = bx(n-1) + \sqrt{1-b^2}\, n_w(n) \quad (9.20)$$

where $n_w(n)$ is Gaussian white noise and b is the parameter between 0 and 1 that determines the level of correlation between adjacent samples of the process x(n).

To identify such a nonlinear Volterra system, the figure 9-2 structure is used in this example. In block A, the 4 by 4 DCT transform matrix is chosen to decorrelate the colored input x(n). H_0, H_1, and H_2 are used in block B. For the 10-memory second-order Wiener model fully expanding equation 7.8, we have a total of 15 $\tilde{Q}$-polynomials, which means that there are 15 coefficients in section C. As in the previous section, properly select scale matrix $\mathbf{S}_{\tilde{Q}}$ as diag$[1, 1, 1, 1, 1, 1/\sqrt{2}, 1/\sqrt{2}, 1/\sqrt{2}, 1/\sqrt{2}, 1/\sqrt{2}, 1/\sqrt{2}, 1, 1, 1, 1, 1, 1]$, which makes the autocorrelation matrix an identity matrix. Without adding plant noise, the simulation results of ensemble averages over 50 independent runs for b = 0.2, 0.5 and 0.9 are shown in figure 9-3. The learning curves of the Volterra model for different b values are shown. It is clear that the Wiener model has better performance. As expected, the nonlinear adaptive Wiener model can identify the Volterra system with colored input. This is because the signal in the DCT transform domain decorrelates the colored input and generates an i.i.d. signal as input to the nonlinear Wiener filter. All $\tilde{Q}$-polynomials in the Wiener model are orthogonal which reduces the eigenvalue spread and increases the convergence speed. Comparing figure 9-3 with figure 9-4, it is clear that if there is no DCT transform, the nonlinear LMS algorithm will not have acceptable performance, especially when b=0.9. The reason is obvious when we check the eigenvalue spread.

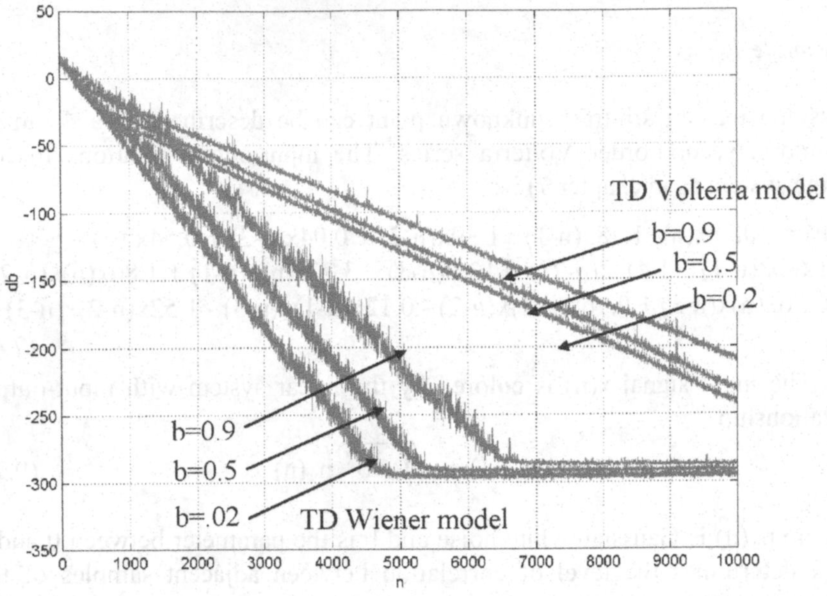

Figure 9-3. Learning curves of transform domain Wiener model and transform domain Volterra model

To examine the eigenvalue spread characteristics, we need to evaluate the eigenvalue of the autocorrelation matrix. For b = 0.5, the autocorrelation matrices of the TD Wiener model and the TD Volterra model are shown in figure 9-5 and figure 9-6 respectively. The eigenvalues are 1 and 45 respectively. The autocorrelation matrix of the corresponding non-TD Wiener model is shown in figure 9-7. As expected, without the whitening block, the autocorrelation matrix shows that there exists a highly correlative relationship between adjacent samples. This results in a large eigenvalue spread.

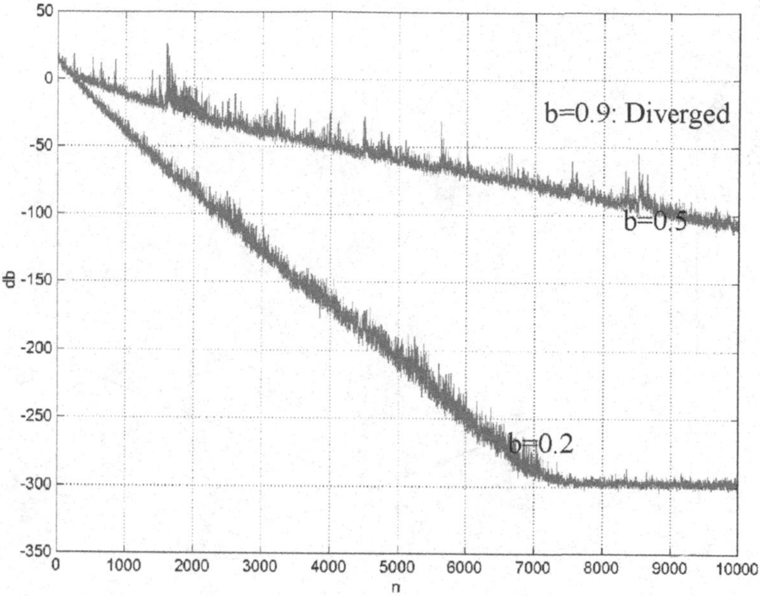

Figure 9-4. Learning curves of non-transform domain Wiener model

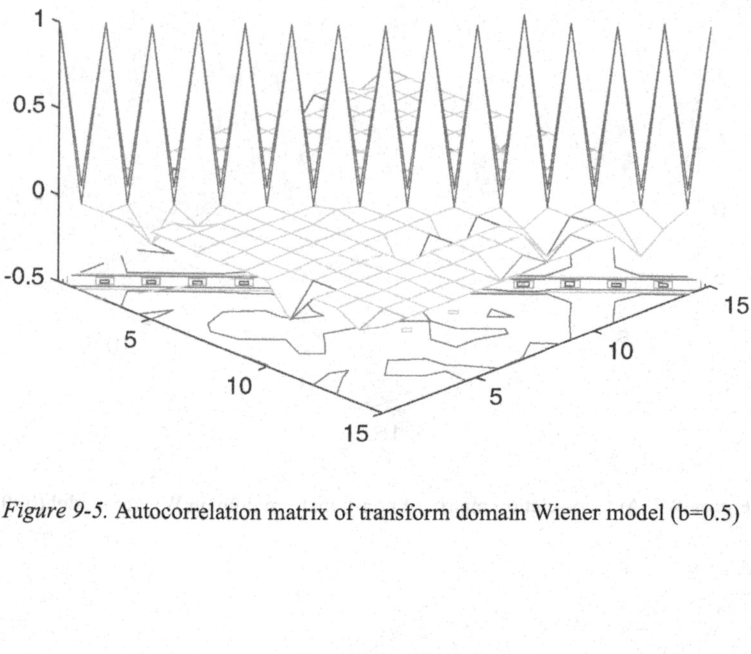

Figure 9-5. Autocorrelation matrix of transform domain Wiener model (b=0.5)

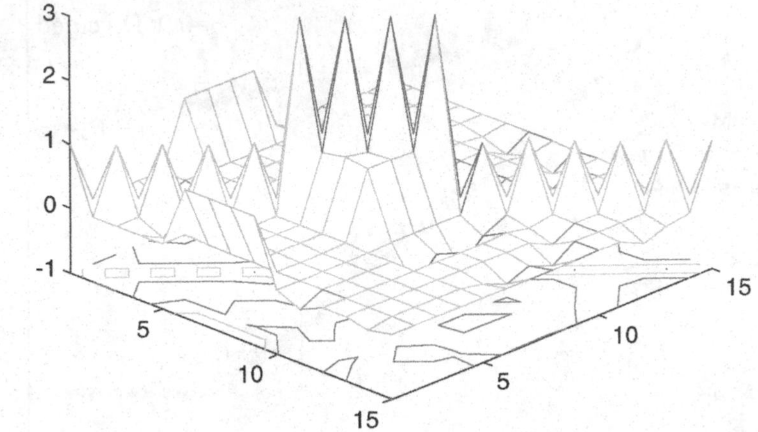

Figure 9-6. Autocorrelation matrix of transform domain Volterra model (b=0.5)

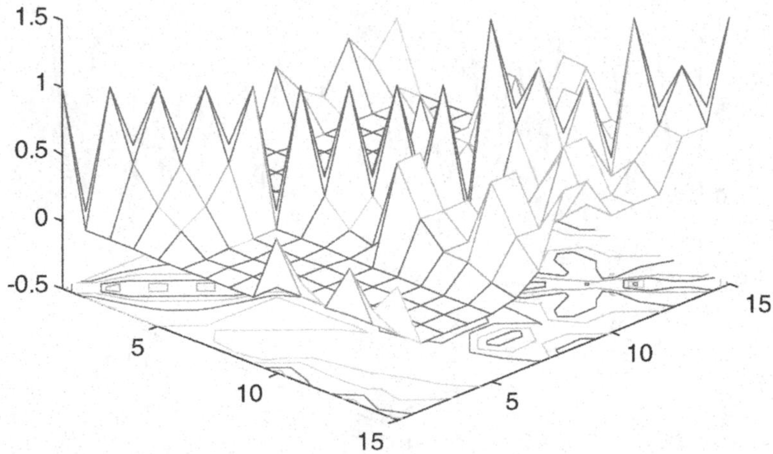

Figure 9-7. Autocorrelation matrix of non-transform domain Wiener model (b=0.5)

9.4 Summary

In this chapter, we have presented alternative algorithms for nonlinear adaptive system identification. We discussed the least-mean-fourth (LMF) version of the algorithm originally presented in chapter 7. The transform domain techniques are applied in order to reduce the eigenvalue spread of the autocorrelation matrix of the input. There are other possible variations of the original algorithms.

In the next chapter, we develop adaptive algorithms based on the recursive-least-squares (RLS)-type algorithm for the nonlinear Wiener model.

Chapter 10

NONLINEAR ADAPTIVE SYSTEM IDENTIFICATION BASED ON WIENER MODELS (PART 4)

Least-squares based algorithms

Introduction

Having developed the nonlinear Wiener LMS-type adaptive filtering algorithms in the previous three chapters, we show in this chapter that the RLS-type algorithm can be applied for the nonlinear Wiener model too. The trade-off is between convergence rate and computational complexity. In addition, for practical VLSI implementation the inverse QR decomposition for the recursive least squares (RLS-type) algorithm can be combined with the nonlinear Wiener model to achieve a highly efficient systolic array architecture.

We consider the feedback loops with the Volterra system, which leads to the recursive Volterra model. The bilinear system (see section 6.2) is an example of such a system (Rugh WJ 2002, Baik 1993). It is hard to analyze recursive Volterra system models by nonlinear Wiener models because the feedback loops are included. Therefore, all the derivations in this chapter are based on the Volterra model. The background of an adaptive nonlinear filter employing nonlinear feedback requires combining QR-decomposition-based algorithms with OLS (orthogonal least squares) algorithms (Billings 1980, Billings 1989). Although the subset selection method (Chen 1989) can be applied to reduce the data matrix size, high computational complexity is expected in general.

10.1 Standard RLS Nonlinear Wiener Adaptive Algorithm

While the nonlinear LMS adaptive filters minimize the instantaneous mean square error by using a gradient search algorithm to estimate the optimum solution for the system to be identified, the RLS adaptive filters, on the other hand, yield the exact solution to the optimization problem by minimizing the square error in a deterministic fashion. By applying a Wiener model, the RLS system minimizes the following cost function:

$$\xi(n) = \sum_{k=1}^{n} \lambda^{n-k} \left(d(k) - \tilde{\mathbf{Q}}^T(k)\mathbf{C}(k) \right)^2 \qquad (10.1)$$

where λ is the forgetting factor ($0<\lambda<1$), which controls the memory span of the adaptive filter. The taps are updated recursively

$$\mathbf{C}(n) = \mathbf{C}(n\text{-}1) + \mathbf{k}(n)e(n), \qquad (10.2)$$

where e(n) is the a priori estimation error which is calculated by

$$e(n) = d(n) - \mathbf{C}^T(n\text{-}1)\tilde{\mathbf{Q}}(n) \qquad (10.3)$$

In equation 10.2, **k**(n) is called the Kalman gain vector which is obtained by (Haykin 1996):

$$\mathbf{k}(n) = \frac{\lambda^{-1}\mathbf{P}(n-1)\tilde{\mathbf{Q}}(n)}{1+\lambda^{-1}\tilde{\mathbf{Q}}^T(n)\mathbf{P}(n-1)\tilde{\mathbf{Q}}(n)} \qquad (10.4)$$

The **P**(n) matrix in equation 10.4 is defined as the inverse matrix of $L\times L$ correlation matrix $\mathbf{R}_{\tilde{Q}\tilde{Q}}(n)$ which is

$$\mathbf{P}(n) = \mathbf{R}_{\tilde{Q}\tilde{Q}}^{-1}(n) = \left(\sum_{i=1}^{n} \lambda^{n-i}\tilde{\mathbf{Q}}(i)\tilde{\mathbf{Q}}^T(i) \right)^{-1} \qquad (10.5)$$

By using the matrix inversion lemma, **P**(n) can be determined recursively by

$$\mathbf{P}(n) = \lambda^{-1}\mathbf{P}(n-1) - \lambda^{-1}\mathbf{k}(n)\tilde{\mathbf{Q}}(n)\mathbf{P}(n-1) \qquad (10.6)$$

The performance of the RLS algorithm does not depend upon the eigen-value spread of $\mathbf{R}_{\tilde{Q}\tilde{Q}}(n)$. The rate of convergence of the RLS algorithm is typically an order of magnitude faster than that of the LMS-type algorithm. This is especially true when the signal-to-measurement noise ratio of the input signals is large. As n approaches infinity, RLS algorithm produces zero misadjustment when operating in a stationary environment. Also, the zero misadjustment property assumes that the exponential weight factor λ equals one; that is, the algorithm operates with infinite memory.

Compared with LMS-type algorithms, the RLS (recursive least square) method provides an alternative approach which has much faster convergence behavior and is independent or less dependent on the statistical properties of the input signal. The price paid for this performance improvement is the increased computational complexity exponentially. In general, for an M-memory and *P*th-order Volterra system, the complexity of the RLS algorithm is $O(M^{2P})$ multiplications per time instant. There are several fast algorithms which have been developed, such as the FTF-like Volterra adaptive filtering algorithm (Lee 1993, 1994) and the lattice structure RLS Volterra adaptive filtering algorithm (Syed 1994). However, most of them are based on the second-order structure, which may not be easy to extend to higher orders because of the highly complicated structures and certain potential numerical instability issues (Mathews 1991). And all the fast algorithms are over-parameterized. This means that, for instance, they require $O(M^3)$ parameters to describe an M-memory second-order Volterra system which has only $O(M^2)$ coefficients.

10.2 Inverse QR Decomposition Nonlinear Wiener Adaptive Algorithm

To obtain higher computational efficiency for RLS filtering, inverse QR decomposition was introduced (Ghirnikar 1992). This method allows the time-recursive least squares weight vector to be updated directly, thus avoiding the highly serial back-substitution in the direct QR algorithm. Additionally, the inverse QR decomposition employs orthogonal rotation operations to recursively update the filter, and thus preserve the inherent stability properties of the direct QR algorithm. Another important benefit of the inverse QR approach is that the rotation computations are easily mapped onto systolic array structures for parallel implementation that can be included in block C of figure 4-2. This is efficient for the VLSI realization of the whole nonlinear Wiener model. For an (L+1)-length filter, the nonlinear Wiener model with inverse QR array is shown in figure 10-1. The r_{ij} and $u_j^{(i)}$ in the array are updated recursively (Alexander 1995):

$$r_{ij}(n) = \lambda^{-1/2} c_i(n) r_{ij}(n-1) - s_i(n) u_j^{(i-1)}(n) \tag{10.7}$$

$$u_j^{(i)}(n) = c_i(n) u_j^{(i-1)}(n) + \lambda^{-1/2} s_i(n) r_{ij}(n-1) \tag{10.8}$$

where $c_i(n)$ and $s_i(n)$ are angle parameters which are

$$s_i(n) = a_i(n) / b_i(n) \quad (10.9)$$

$$c_i(n) = b^{(i-1)}(n) / b^{(i)}(n) \quad (10.10)$$

The $a_i(n)$ and $b_i(n)$ can be determined recursively too:

$$a_i(n) = \lambda^{-1/2} \sum_{j=1}^{i} r_{ij}(n-1)\tilde{\mathbf{Q}}(n-j+1) \quad (10.11)$$

$$b^{(i)}(n) = \sqrt{[b^{(i-1)}(n)]^2 + a_i^2(n)} \quad (10.12)$$

The coefficients are updated according to:

$$w_k(n) = w_k(n-1) + z(n)u_k^{(L)}(n), \text{ for } k = 0, 1, ..., L \quad (10.13)$$

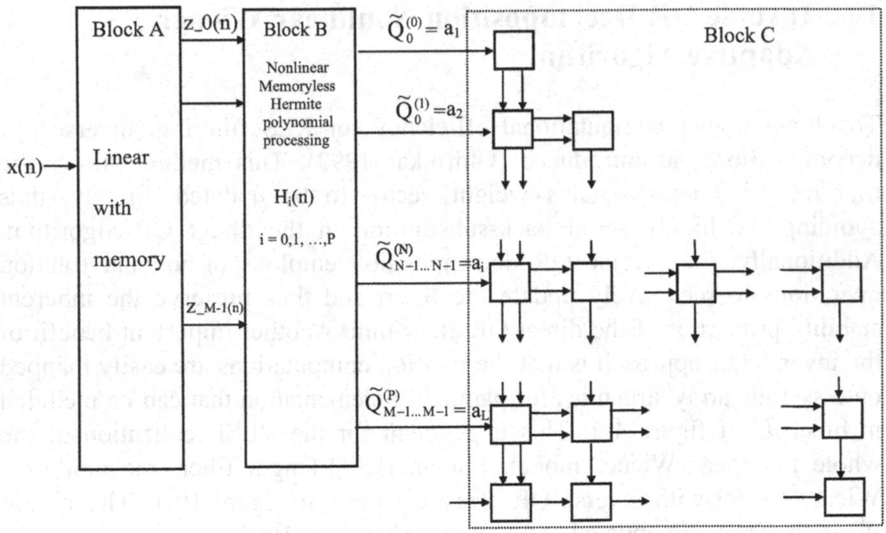

Figure 10-1. Nonlinear Wiener model with inverse QR array

where z(n) is defined as

$$z(n) = e(n \mid n-1) / b^{(L)}(n) \tag{10.14}$$

where the prediction error $e(n|n\text{-}1) = d(n)\text{-}\mathbf{C}^T(n\text{-}1)\tilde{\mathbf{Q}}(n)$. This systolic array for parallel implementation of inverse QR algorithm after initialization produces L+1 new filter coefficients every L+2 clock cycles without back substitution. Therefore, the architecture in figure 10-1 is a very efficient and practical way to realize the RLS algorithm.

10.3 Recursive OLS Volterra Adaptive Filtering

All the nonlinear adaptive filtering algorithms presented till now are based on the Wiener model, which can be used to identify the truncated Volterra systems. However, difficulties arise if the Volterra system involves the feedback loops. For more general consideration, this section discusses the QR-based recursive OLS (orthogonal least square) algorithm for recursive Volterra adaptation. This is the general extension version of the adaptive bilinear system. Combining with the subset selection algorithm, the size of the data matrix may be reduced, which may help to decrease the computational complexity. All derivations in this section are based on the Volterra model.

10.3.1 Recursive OLS QR Decomposition Adaptive Algorithm

Based on the NARMAX (nonlinear autoregressive moving average with exogenous input) model (Billings 1989, Chen 1989), consider an r-inputs and m-outputs discrete-time multivariable non-linear stochastic system:

$$\mathbf{y}(n) = f[\mathbf{y}(n\text{-}1), ..., \mathbf{y}(n\text{-}n_y), \mathbf{u}(n\text{-}1), ..., \mathbf{u}(n\text{-}n_u), \mathbf{e}(n\text{-}1), ..., \mathbf{e}(n\text{-}n_e)] + \mathbf{e}(n) \tag{10.15}$$

where

$$\mathbf{y}(n) = [\, y_1(n), y_2(n), ..., y_m(n) \,]^T,$$
$$\mathbf{u}(n) = [\, u_1(n), u_2(n), ..., u_r(n) \,]^T,$$
$$\mathbf{e}(n) = [\, e_1(n), e_2(n), ..., e_m(n) \,]^T$$

are the system output, input, and noise respectively; n_y, n_u and n_e are the maximum lags in the output, input, and noise; $\mathbf{e}(n)$ is a zero mean independent sequence; and f(.) is some vector-valued nonlinear function. The block diagram of equation 10.17 is shown in figure 10-2.

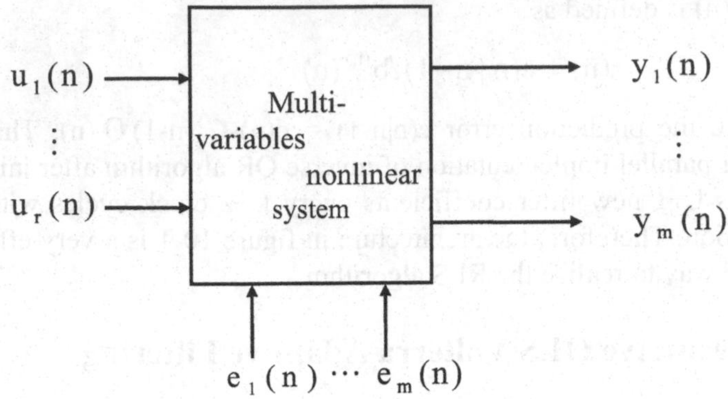

Figure 10-2. Block diagram for NARMAX model

The scalar form of equation 10.14 can be decomposed into m equations as follows:

$$y_i(n) = f_i(y_1(n-1),\ldots,y_1(n-n_y),\ldots,y_m(n-1),\ldots,y_m(n-n_y),$$
$$u_1(n-1),\ldots,u_1(n-n_u),\ldots,u_r(n-1),\ldots,u_r(n-n_x),$$
$$e_1(n-1),\ldots,e_1(n-n_u),\ldots,e_m(n-1),\ldots,e_m(n-n_e)) + e_i(n), i = 1, ..., m \quad (10.16)$$

For simplification and without lost of generality, a special case of NARX (nonlinear autoregression with exogenous input) model is considered here which can be written as:

$$\mathbf{y}(n) = f(\mathbf{y}(n\text{-}1), ..., \mathbf{y}(n\text{-}n_y), \mathbf{u}(n\text{-}1), ..., \mathbf{u}(n\text{-}n_x)) + \mathbf{e}(n) \quad (10.17)$$

The scalar form of equation 10.19 is:

$$y_i(n) = f_i(y_1(n-1),..., y_1(n-n_y),..., y_m(n-1),..., y_m(n-n_y),$$
$$u_1(n-1),..., u_1(n-n_x),..., u_r(n-1),..., u_r(n-n_x)) + e_i(n), \quad i = 1,\ldots, m \quad (10.18)$$

In reality, the non-linear form of $f_i(.)$ in equation 10.18 is generally unknown. Any discrete function $f_i(.)$, however, can be arbitrarily well approximated by polynomial models, expanding $f_i(.)$ as a first-order recursive Volterra series which can be written as follows:

$$y_i(n) = h_0^{(i)} + \sum_{k_1=1}^{M} h_{k_1}^{(i)} x_{k_1}(n) + \sum_{k_1=1}^{M}\sum_{k_2=k_1}^{M} h_{k_1 k_2}^{(i)} x_{k_1}(n) x_{k_2}(n) +$$

$$\ldots + \sum_{k_1=1}^{M} \ldots \sum_{k_M=k_{M-1}}^{M} h_{k_1 \ldots k_M}^{(i)} x_{k_1}(n) \ldots x_{k_M}(n) + e_i(n) \qquad (10.19)$$

where $M = m \times n_y + r \times n_u$ and

$$x_1(n) = y_1(n-1),\ x_2(n) = y_1(n-2),\ \ldots,\ x_{m \times n_y}(n) = y_m(n-n_y),$$
$$x_{m \times n_y + 1}(n) = u_1(n-1),\ \ldots,\ x_M(n) = u_r(n-n_u)$$

It is clear that each subsystem model in equation 10.18 belongs to the linear regression model.

Define

$$d_i(n) = \sum_{j=1}^{M} a_{ij}(n) h_j(n) + \xi_i(n), \qquad i = 1, \ldots, N \qquad (10.20)$$

where N is the desired data length and the $a_{ij, j=1,\ldots,M}(n)$ are $x_1(n)$ to $x_M(n)$ with all the linear and nonlinear combinations up to the order M for each fixed index i, ξ_i is some modeling error, and the $h_{j=1,\ldots,M}(n)$ are the unknown parameters to be estimated. In linear regression analysis, $a_{ij}(n)$ are often referred to as predictors. The matrix form of equation 10.20 can be written as:

$$\mathbf{d}(n) = \mathbf{A}(n)\mathbf{H}(n) + \mathbf{E}(n) \qquad (10.21)$$

where

$$\mathbf{d}(n) = \begin{bmatrix} d_1(n) \\ \vdots \\ d_N(n) \end{bmatrix}, \qquad \mathbf{A}(n) = [\mathbf{a}_1(n) \ \cdots \ \mathbf{a}_M(n)],$$

$$\mathbf{H}(n) = \begin{bmatrix} h_1(n) \\ \vdots \\ h_M(n) \end{bmatrix}, \ \mathbf{E}(n) = \begin{bmatrix} \xi_1(n) \\ \vdots \\ \xi_N(n) \end{bmatrix}$$

and

$$\mathbf{a}_j(n) = \begin{bmatrix} a_{1j}(n) \\ \vdots \\ a_{Nj}(n) \end{bmatrix}, \quad j = 1, \ldots, M$$

If the significant terms are known a priori and only they are used for the regression matrix $\mathbf{A}(n)$, a linear least square (LS) problem can be defined as follows:

Find the parameter estimate $\hat{\mathbf{H}}(n)$ which minimizes $\|\mathbf{d}(n)-\mathbf{A}(n)\mathbf{H}(n)\|$ where $\|\cdot\|$ is the Euclidean norm. There are several methods to solve this LS problem (Haykin 1996). Here, we use QR decomposition based on the Givens rotation method, which can be implemented recursively with good stability property and save a lot of computation memory. Assume $N\times M$ $\mathbf{A}(n)$ matrix with $N = M+1$; equation 10.21 can be written as:

$$\mathbf{E}(n) = \mathbf{d}(n) - \mathbf{A}(n)\mathbf{H}(n) \tag{10.22}$$

We can find a series of $N\times N$ Givens rotations matrices represented by $\mathbf{T}$, such that:

$$\mathbf{T}(n)\mathbf{E}(n) = \mathbf{d}'(n) - \mathbf{A}'(n)\mathbf{H}(n) = \begin{pmatrix} \mathbf{p}(n) \\ v(n) \end{pmatrix} - \begin{pmatrix} \mathbf{R}(n) \\ \mathbf{0}^T \end{pmatrix}\mathbf{H}(n) \tag{10.23}$$

where $\mathbf{p}(n)$ is $M\times 1$ column vector, $v(n)$ is a scalar number, $\mathbf{R}(n)$ is an upper triangular $M\times M$ matrix, and $\mathbf{0}^T$ is a $1\times M$ row vector. The LS solution $\hat{\mathbf{H}}(n)$ for equation 10.23 can be determined by solving

$$\hat{\mathbf{H}}(n) = \mathbf{R}^{-1}(n)\mathbf{p}(n) \tag{10.24}$$

using backward substitution. If we normalize each row of $\mathbf{R}(n)$ and $\mathbf{p}(n)$ with respect to the diagonal elements of $\mathbf{R}(n)$, we can get normalized $\mathbf{R}_N(n)$ and $\mathbf{p}_N(n)$. The $\mathbf{R}_N(n)$ is an upper triangular matrix with all 1's in its diagonal, which is:

$$\mathbf{R}_N(n) = \begin{bmatrix} 1 & r_{12} & r_{13} & \cdots & r_{1M} \\ & 1 & r_{23} & \cdots & r_{2M} \\ & & \ddots & \ddots & \vdots \\ & & & 1 & r_{M-1,M} \\ & & & & 1 \end{bmatrix} \tag{10.25}$$

Therefore, the LS solution of $\mathbf{H}(n)$ can be rewritten as:

$$\hat{\mathbf{H}}(n) = \mathbf{R}_N^{-1}(n)\mathbf{p}_N(n) \tag{10.26}$$

The relation between $\mathbf{R}_N(n)$ and $\mathbf{A}(n)$ is defined by $\mathbf{Q}(n)$:

$$\mathbf{A}(n) = \mathbf{Q}(n)\mathbf{R}_N(n) \tag{10.27}$$

where $\mathbf{Q}(n)$ is an $N\times M$ matrix which can be found by:

$$\mathbf{Q}(n) = \mathbf{A}(n)\mathbf{R}_N^{-1}(n) \tag{10.28}$$

Because $\mathbf{R}_N(n)$ is an upper triangular matrix, its inverse can be calculated as (Benesty 1992):

$$\mathbf{R}_N^{-1}(n) = \mathbf{G}_1\mathbf{G}_2 \ldots \mathbf{G}_{M-1} \tag{10.29}$$

where

$$\mathbf{G}_i(n) = \begin{pmatrix} 1 & 0 & 0 & 0 & 0 & \cdots & 0 \\ & 1 & 0 & 0 & 0 & \cdots & \\ & & \ddots & & & \cdots & \\ & & & 1 & -r_{i+1,i+2} & \cdots & -r_{i+1,M} \\ & & & & \ddots & & \\ & & & & & 1 & 0 \\ & & & & & & 1 \end{pmatrix} \leftarrow \text{(i+1)row} \tag{10.30}$$

The $\mathbf{Q}(n)$ matrix can be written as:

$$\mathbf{Q}(n) = [\mathbf{q}_1 \quad \cdots \quad \mathbf{q}_M] \tag{10.31}$$

We note that all the columns of $\mathbf{Q}(n)$ are mutually orthogonal. Therefore, equation 10.31 satisfies

$$\mathbf{Q}^T(n)\mathbf{Q}(n) = \mathbf{D}(n) \tag{10.32}$$

where $\mathbf{D}(n)$ is the positive diagonal matrix. Substituting equation 10.30 in equation 10.24, the relation of $\mathbf{E}(n)$, $\mathbf{Q}(n)$ and $\mathbf{p}_N(n)$ is:

$$\hat{\mathbf{E}}(n) = \mathbf{d}(n) - \mathbf{Q}(n)\mathbf{R}_N(n)\mathbf{H}(n) = \mathbf{d}(n) - \mathbf{Q}(n)\mathbf{p}_N(n) \tag{10.33}$$

To form $\mathbf{A}(n+1)$ and $\mathbf{d}(n+1)$ for the next iteration, we need to append new data $\mathbf{x}^T(n+1)$ and $d(n+1)$ at the end of $\mathbf{D}^{1/2}\mathbf{R}_N(n)$ and $\mathbf{d}(n+1)$, which are:

$$\mathbf{A}(n+1) = \begin{pmatrix} \mathbf{D}^{1/2}\mathbf{R}_N(n) \\ \mathbf{x}^T(n+1) \end{pmatrix} \tag{10.34}$$

$$\mathbf{d}(n+1) = \begin{pmatrix} \mathbf{D}^{1/2}\mathbf{p}_N(n) \\ d(n+1) \end{pmatrix} \tag{10.35}$$

Now we can go back to equation 10.23 to start the next iteration.

10.3.2 Subset Selection

The number of columns in $\mathbf{A}(n)$ is sometimes very large, perhaps in the thousands. It is shown that a simple and efficient subset selection algorithm can be applied that determines $\mathbf{A}_s(n)$, the subset of $\mathbf{A}(n)$. This is done using a GramSchmidt procedure in an embedded forward-regression manner by continuously choosing columns of $\mathbf{A}(n)$ for which the sum of squares of residual error is reduced to a desired tolerance. The detailed procedure of subset selection can be found in (Chen 1989) and (Billings 1989).

10.4 Computer Simulation Examples

Here we present a few examples of computer simulation for the algorithms described in this and in some previous chapters.

10.4.1 Nonlinear RLS Wiener Adaptive Algorithm

Example 1:

For faster convergence, we can use the RLS-type algorithm with the nonlinear Wiener model which was described in section 10.1. To verify the theory, reconsider example 1 given in section 7.2. Set SNR = -40 db. The forgetting factor is chosen to be 0.995. The learning curve that has been averaged over 50 runs is shown in figure 10-3. The superior convergence behavior of the RLS algorithm in this example is obvious. The price paid for the better performance is that the computational complexity is $O(M^4)$, which is considerably more than the LMS adaptive filters of $O(M^2)$. For practical VLSI implementation, the systolic structure in figure 10-1 can be used. Simulation was performed using equations 10.7 through 10.14. As expected, the learning curves for SNR = 40 and SNR = 20 shown in figure 10-4 have a similar performance as the RLS algorithm. And the throughput of the array is every 7 clock cycles for 6 filter coefficients.

10.4.2 Recursive Volterra Adaptive Filter Algorithm

Example 2:

Consider a system identification problem as in the previous subsection. The unknown system is a Volterra polynomial equation:

$$y(n) = 0.15y(n-1) - 0.3y(n-2) + 0.8x(n-1) - 0.52x^2(n-2) + 0.215x(n-1)x(n-2) \tag{10.36}$$

where the system input x(n) is a uniform distribution independent sequence with zero mean and unit variance. The noise is zero mean with variance 10^{-4}.

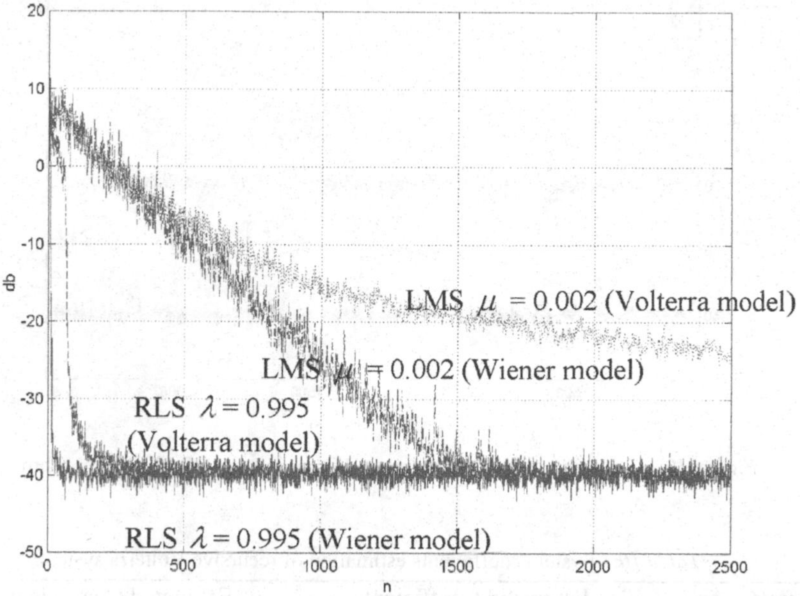

Figure 10-3. Learning curves of nonlinear LMS and RLS algorithms

This is the polynomial NARX model with $l = n_y = n_x = 2$. The full model set contains 15 terms. Using the iterative scheme discussed in chapter 5, with 1500 data input, instead of storing all the data in one matrix, the maximum dimension for matrix $\mathbf{A}'(n)$ is only 16×15. Through 50 runs, the mean square error learning curve is shown in figure 10-5. The final model coefficients are shown in table 10.1. Applying subset selection technique in section 10.3.2 by setting $\rho = 10^{-5}$, the estimated coefficients are shown in table 10.1 as well. The mean square error learning curve is shown in figure 10-6. Basically both curves are similar. But table 10.1 shows that the latter one has better accuracy. The trade-off is that the subset selection procedure increases the computational complexity and processing time.

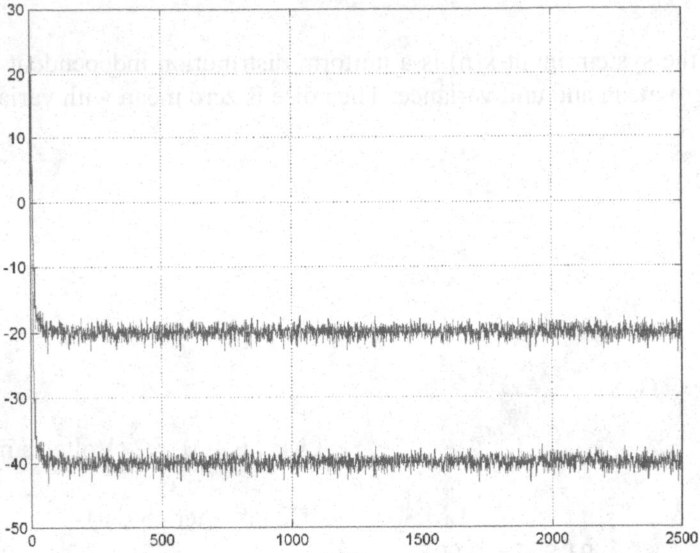

Figure 10-4. Learning curves of nonlinear Wiener inverse QR (IQR) algorithm

Table 10.1 System coefficients estimation of recursive Volterra system

Terms	Estimated Coefficients without Subset Selection	Estimated Coefficients with Subset Selection
1	0.00000611	0
y(n-1)	0.15017880	0.14993425
y(n-2)	-0.29992202	-0.29996695
u(n-1)	0.79992230	0.79997514
u(n-2)	0.00015261	0
y^2(n-1)	0.00011170	0
y^2(n-2)	0.00005327	0
u^2(n-1)	-0.00001002	0
u^2(n-2)	-0.52009616	-0.52002455
y(n-1)y(n-2)	0.00003872	0
y(n-1)u(n-1)	-0.00016826	0
y(n-1)u(n-2)	-0.00000482	0
y(n-2)u(n-1)	0.00005996	0
y(n-2)u(n-2)	-0.00006095	0
u(n-1)u(n-2)	0.21516695	0.21503372

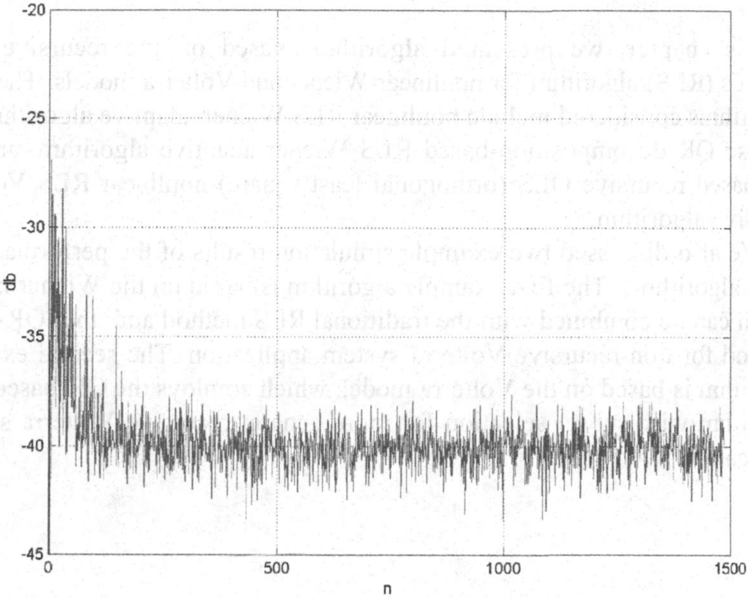

Figure 10-5. Learning curve of recursive Volterra adaptive filter algorithm

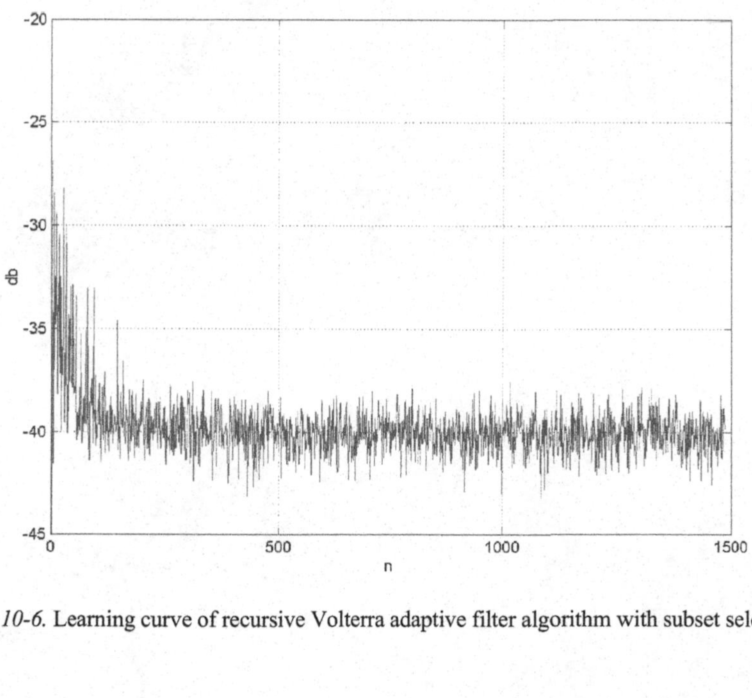

Figure 10-6. Learning curve of recursive Volterra adaptive filter algorithm with subset selection

10.5 Summary

In this chapter, we presented algorithms based on the recursive least squares (RLS) algorithm for nonlinear Wiener and Volterra models. The three algorithms considered include nonlinear RLS Wiener adaptive algorithm, the inverse QR decomposition-based RLS Wiener adaptive algorithm, and the QR-based recursive OLS (orthogonal least square) nonlinear RLS Volterra adaptive algorithm.

We also discussed two example simulation results of the performance of these algorithms. The first example algorithm is based on the Wiener model, which can be combined with the traditional RLS method and the IQR-based method for non-recursive Volterra system application. The second example algorithm is based on the Volterra model, which employs the QR-based OLS algorithm with subset selection for more general recursive Volterra system application.

Chapter 11

CONCLUSIONS, RECENT RESULTS, AND NEW DIRECTIONS

Summary

In this book, we have presented an introduction to the area of nonlinear adaptive system identification that builds upon an introduction to polynomial nonlinear systems and adaptive signal processing.

The area of nonlinear adaptive system identification has found many applications in control, communications, biological signal processing, image processing, etc.

In the first of the three parts of the book is some useful introductory material on important background information, including nonlinear systems, polynomial modeling of nonlinear systems, Volterra and Wiener models for nonlinear systems, and an introduction to system identification and adaptive filtering. In the second part, we describe the different stochastic gradient-type algorithms for nonlinear system identification methods, first based on the Volterra model, then later on the Wiener model. The final part describes the recursive least-squares-type algorithms for nonlinear systems.

Based on the self-orthogonal property of Hermite polynomials, we developed a delay-line version of the discrete nonlinear Wiener model. This structure is very suitable for LMS-type adaptive algorithms, especially for the Gaussian white input.

Adaptive algorithms based on the discrete nonlinear Wiener model offer the following advantages over those based on the discrete Volterra model:

- faster convergence,
- less number of computations (no over-parameterization),

- applicability to non-Gaussian and non-white signal environments, and
- wide applicability to practical problems due to the possibility of real-time implementations.

11.1 Conclusions

The methods and algorithms presented in this book were based on the discrete-time nonlinear Wiener and Volterra models applied to adaptive system identification. The algorithms were developed mostly for second- and third-order models but can be extended to higher-order models.

All the algorithms and performance results were verified and confirmed by computer simulations. To illustrate practical real-time implementations, some of the methods presented in this book have been implemented on DSP processors (Ryan 1999, Ogunfunmi 1998).

Alternatively, the discrete nonlinear Wiener model can also be used with RLS-type adaptive algorithms. For VLSI implementation consideration, the inverse QR nonlinear Wiener adaptive structure is investigated. Although there is no over-parameterization issue and it has very fast convergence speed, the computational complexity is considerably higher than with the LMS-type methods.

For a general IIR Volterra system, based on QR decomposition, we presented the recursive OLS algorithm. This is a high computational complexity structure. Therefore the subset selection is needed to reduce the data matrix size.

11.2 Recent Results and New Directions

We have focused on Volterra and Wiener models. There are other models, such as Hammerstein models, which are useful and need more research. Recent research results indicate a good promise of practical applications for the adaptive algorithms developed based on Hammerstein models (Jeraj 2005, Jeraj 2006a, Jeraj 2006b, Voros 2005).

Another area of possible research work is the development of "fast" algorithms based on the Wiener model. Some "fast" algorithms have been developed for the Volterra model; however, because the Wiener model has clear advantages over the Volterra model, its advantages will likely carry over to the "fast" algorithm development. In addition, the block-LMS and frequency-domain algorithms for nonlinear systems based on the nonlinear Wiener model can be explored to reduce the computational complexity further.

We have presented convergence analysis of the algorithms here based on the so-called "independence assumptions." A possible direction will be to investigate the convergence analysis without those "independence assumptions."

Many of the Volterra model-based applications are limited to second-order or at most third-order models. This is because of the exponential increase in the number of parameters and the resulting complexity of the algorithms. There are, however, some applications which require more than third-order Volterra or Wiener models. For these applications, it is useful to develop new adaptive algorithms which will converge and be practical.

Neural network models have been used recently in the system identification of nonlinear systems (Billings 2005).

REFERENCES

(Alexander 1995) Alexander, S. Thomas and Ghirnikar, Avinash L., "A Method for Recursive Least Squares Filtering Based Upon Inverse QR Decomposition," *IEEE Trans. Signal Processing,* vol. 41, no. 1, pp. 20–30, Jan. 1995.

(Baik 1993) Baik, H. K. and Mathews, V. J., "Adaptive bilinear lattice filters," *IEEE Trans. Signal Processing,* vol. 41, no. 6, June 1993, pp. 2033–2046.

(Beckmann 1973) Beckmann, P., Orthogonal Polynomials for Engineers, Golem Press, Boulder, Colorado, 1973.

(Benedetto 1979) Benedetto, S., Biglieri, E. and Daffara, R., "Modeling and performance evaluation of nonlinear satellite links - A Volterra series approach," *IEEE Trans. Aerospace and Electronic Systems,* vol. AES-15, pp. 494–507, 1979.

(Benedetto 1983) Benedetto, S. and Biglieri, E., "Nonlinear equalization of digital satellite channels," *IEEE J. Selected Areas on Commun.,* vol. SAC-1, pp. 57–62, 1983.

(Benesty 1992) Benesty, Jacob and Duhamel, Pierre, " A Fast Exact Least Mean Square Adaptive Algorithm," *IEEE Trans. Signal Processing,* vol. 40, no. 12, Dec. 1992, pp. 2904–2920.

(Bershad 1999) Bershad, Neil J., Celka, Partric and Vesin, Jean Maarc, "Stochastic Analysis of Gradient Adaptive Identification of Nonlinear Systems with Memory for Gaussian Data and Noisy Input and Output Measurements," *IEEE Trans. Signal Processing,* vol. 47, no. 3, pp. 675–689, Mar. 1999.

(Biglieri 1984) Biglieri, E., Gersho, A., Gillin, R. D. and Lim, T. L., "Adaptive cancellation of nonlinear interference for voiceband dada transmission," *IEEE J. Select. Area Commun.,* vol. SAC-2, pp. 765–777, Sept. 1984.

(Billings 1980) Billings, S. A., "Identification of nonlinear systems: Asurvey" *IEE Proceedings,* Pt. D, vol. 127, no. 6, pp. 272–285, Nov. 1980.

(Billings 1984) Billings, S. A. and Voon, W. S. F., "Least squares-parameter estimation algorithms for nonlinear systems,' *Int. J. System Sci.,* Vol. 15, No. 6, pp. 601–615, 1984.

(Billings 1989) Billings, S. A., Chen, S. and Korenberg, M. J., "Identification of MIMO non-linear systems using a forward-regression orthogonal estimator," *Int. J. control,* 1989, vol. 49, no. 6, pp. 2157–2189.

(Billings 2005) Billings, S. A. and Wei, Hua-Liang, "A New Class of Wavelet Networks for Nonlinear System Identifications" *IEEE Transactions on Neural Networks,* vol. 16, no. 4, pp. 272–285, July 2005.

(Brogam 1991) Brogam, William L., Moderm Control Theory, Third edition, Prentice Hall, N. J., 1991.

(Broshtein 1985) Broshtein, I. N. and Semendyayev, K.A., *Handbook of Mathematics,* VNR., New York, 1985.

(Chang 1997a) Chang, Shue-Lee and Ogunfunmi, Tokunbo, "Performance analysis of nonlinear adaptive filter based on LMS algorithm," *Proc. 31th ASILOMAR Conf.,* Monterey, California, 1997.

(Chang 1997b) Chang, Shue-Lee and Ogunfunmi, Tokunbo, "Recursive Orthogonal Least Square Method and its application in nonlinear adaptive filtering," *Proc. MILCOM Conf.,* Monterey, California, 1997.

(Chang 1998a) Chang, Shue-Lee and Ogunfunmi, Tokunbo, "LMS/LMF and QR Volterra System Identification based on Nonlinear Wiener Model," *Proc. IEEE ISCAS,* Monterey, California, 1998.

(Chang 1998b) Chang, Shue-Lee and Ogunfunmi, Tokunbo, "An Improved Nonlinear Adaptive Filter and PDF Estimator using discrete Fourier Transform," *Proc. IEEE DSP Workshop,* Bryce Canyon, Utah, 1998.

(Chang 1998c) Chang, Shue-Lee and Ogunfunmi, Tokunbo, "Volterra adaptive system identification based on nonlinear wiener model," *Proc. IASTED,* Las Vegas, Nevada, 1998.

(Chang 1999) Chang, Shue-Lee and Ogunfunmi, Tokunbo, "Performance Analysis of Third-order Nonlinear Wiener Adaptive systems," *Proc. IEEE ICASSP,* Phoenix, Arizona, 1999.

(Chang 2000a) Chang, Shue-Lee and Ogunfunmi, Tokunbo, "Transform Domain Second-order Nonlinear Adaptive Filtering for Colored Gaussian Signals," *Proceedings of the IEEE Workshop on Signal Processing Implementation Systems* (SiPS), Lafayette, Louisiana, 2000.

(Chang 2003) Chang, Shue-Lee and Ogunfunmi, Tokunbo, "Stochastic Gradient Based Third-order Volterra System Identification by Using Nonlinear Wiener Adaptive Algorithm," *IEE Proceedings-Vision, Image and Signal Processing,* vol. 150, no. 2, pp. 90–98, April 2003.

(Chen 1989) Chen, S., Billings, S. A. and Luo, W., "Orthogonal least squares methods and their application to non-linear system identification," *Int. J. control,* 1989, vol. 50, no. 5, pp. 1873–1896.

(Diaz 1988) Diaz, H. and Desrochers, A. A. "Modeling of nonlinear-discrete-time systems from input-output data," *Automatica,* vol. 24, no. 5, pp. 629–641, 1988

(Diniz 2002) Diniz, Paulo, Adaptive Filtering: Algorithms and Practical Implementation, Springer Publishers, Second edition, 2002.

(Efunda 2006) Engineering Fundamentals, Orthogonal Polynomials, URL: http://www.efunda.com/math/hermite/index.cfm

(Fejzo 1995) Fejzo, Zoran and Lev-Avi, Hanoch, "Adaptive nonlinear Wiener-Laguerre-Lattice models," *Proceedings of IEEE ICASSP,* pp. 977–980, 1995.

(Fejzo 1997) Fejzo, Zoran and Lev-Avi, Hanoch, "Adaptive Laguerre-Lattice Filters," *IEEE Tran. Signal Processing,* vol. 45, pp. 3006–3016, Dec. 1997.

(Ghirnikar 1992) Ghirnikar, Avinash L., Alexander, S. T. and Plemmons, R. J., "A Parallel Implementation of the Inverse QR Adaptive Filter," *Computer Elect. Eng.,* vol. 18, no. 3–4, pp.192–200, 1992.

(Gill 1981) Gill, Philip, Murray, Walter and Wright, Margaret, Practical Optimization, Academic Press, London, 1981.

(Hashad 1994) Hashad, Atalla, "Analysis of non-Gaussian processes using the Wiener model of Discrete Nonlinear Systems," *Ph.D dissertation,* Dept of Electrical Engineering Naval Postgraduate School, 1994.

(Haykin 1996) Haykin, Simon, Adaptive Filter Theory, Prentice-Hall, Inc., Englewood, N. J., Third edition, 1996.

(Haykin 1998) Haykin, Simon, Neural Networks: A Comprehensive Foundation (2nd. edition), Prentice-Hall, Inc., Englewood, N. J., Third edition, 1998.

(Haber 1985) Haber, R. "Nonlinearity Tests for Dynamic Processes," *IFAC* Symposium on Identification and System Parameter Estimation, pp. 409–414, York, UK, 1985.

(Hush 1998) Don Hush, "Nonlinear Signal Processing Methods," *IEEE Signal Processing Magazine,* pp. 20–22, May 1998.

(Ibnkahla 2003) M. Ibnkahla, "Nonlinear Channel Identification Using-Natural Gradient Descent: Application to Modeling and Tracking,"Chapter in *Soft Computing for Communications,* Edited by L.Wang, Springer-Verlag, 2003.

(Ibnkahla 2002) M. Ibnkahla, "Statistical Analysis of Neural Modelingand Identification of Nonlinear Systems with Memory," *IEEE Trans. Signal Processing,* vol. 50, pp. 1508–1517, June 2002.

(Im 1996) Im, Sungbin and Powers, Edward J., "A Fast Method of Discrete Third-order Volterra Filtering," *IEEE Trans. Signal Processing,* vol. 44, no. 9, pp. 2195–2208, Sept. 1996.

(Jeraj 2005) Jeraj, Janez, "Adaptive Estimation and Equalization of Nonlinear Systems," PhD Dissertation, Dept. of Electrical and Computer Engineering, University of Utah, Salt Lake City, Utah, 2005.

(Jeraj 2006a) Jeraj, Janez and Mathews, V. John, "A Stable Adaptive Hammerstein Filter Employing Partial Orthogonalization of the Input Signals," *IEEE Trans. Signal Processing,* vol. 54, no. 4, pp. 1412–1420, April 2006.

(Jeraj 2006b) Jeraj, Janez and Mathews, V. John, "Stochastic Mean-Square Performance Analysis of an Adaptive Hammerstein Filter," *IEEE Trans. Signal Processing,* vol. 54, no. 6, pp. 2168–2177, June 2006.

(Kailath 1979) Kailath, Thomas, *Linear Systems,* Prentice-Hall., New York, 1979.

(King 1977) King, R. E. and Paraskevopoulos, P. N., "Digital Laguerre filters," *Int. J. Circ. Theory Applicat.,* vol. 5, pp. 81–91, 1977.

(Koh 1983) Koh, T. and Powers E. J., "An adaptive nonlinear filter with lattice orthogonalization," *Proc. IEEE Conf. Acoust., Speech, Signal Proc.,* Boston, Massachusetts, April 1983, pp. 37–39.

(Koh 1985) Koh, T. and Powers, E. J., "Second-order Volterra filtering and its application to nonlinear system identification," *IEEE Trans. Acoust., speech, Signal Proc.,* vol. ASSP-33, no. 6, pp. 1445–1455, Dec. 1985.

(Korenberg 1991) Korenberg, M. J., "Orthogonal approaches to time-series analysis and system identification," *IEEE Signal Processing Mag.,* vol. 8, pp. 29–43, 1991.

(Lathi 2004), Lathi, B.P., Linear Systems and Signals, Second edition, Oxford University Press, 2004.

(Lathi 2000), Lathi, B.P., Signal Processing and Linear Systems, Oxford University Press, 2000.

(Lee 1993) Lee, Junghsi and Mathews, V. John, "A Fast Recursive Least Square Adaptive Second-Order Volterra Filter and Its Performance Analysis," *IEEE Trans. Signal Processing,* vol. 41, no. 3, Mar. 1993, pp. 1087–1102.

(Lee 1994) Lee, Junghsi and Mathews, V. John, "Adaptive Bilinear Predictors," *Proc., IEEE ICASSP,* Adelaide, Australia, April 1994.

(Li 1996) Li, X., Jenkins, W. K. and Therrien, C. W., "Algorithms for Improved Performance in Adaptive Polynomial Filter with Gaussian Input Signals," *Proc. of The 30th Asilomar Conf. Record on Signal, Systems & Computers,* Nov. 1996, pp. 267–270.

(Li 1998) Li, X., Jenkins, W. K. and Therrien, C. W., "Computationally Efficient Algorithms for Third Order Adaptive Volterra Filters," *Proc. of IEEE ICASSP,* May 1998, pp. 1405–1408.

(Ljung 1999) Lennart Ljung, *System Identification: Theory for the User,* Second Edition, Prentice-Hall, New Jersey, 1999.

(Marmarelis 1993) Marmarelis, "Identification of Nonlinear Biogical Systems Using Laguerre Expansion of Kernel," *Annals of Biomedical Engineering,* vol. 21, pp. 573–589, 1993.

(Marmarelis 1978) Marmarelis, P. Z. and Marmarelis, V. Z., *Analysis of Physiological Systems,* Plenum Press, 1978.

(Mathews 1991) Mathews, V. John, "Adaptive Polynomial Filters," *IEEE Signal Processing Mag.,* vol. 8, no. 3, pp. 10–26, July 1991.

(Mathews 1995) Mathews, V. John, "Orthogonalization of Correlated Gaussian Signals for Volterra System Identification," *IEEE Signal Processing Letters,* vol. 2, no. 10, Oct. 1995, pp. 188–190.

(Mathews 1996) Mathews, V. John, "Adaptive Volterra Filters Using Orthogonal Structure," *IEEE Signal Processing Letters,* vol. 3, no. 12, Dec. 1996, pp. 307–309.

(Mathews 2000) Mathews, V. John and Sicuranza, Giovanni L., Polynomial Signal Processing, Wiley and Sons, New York, 2000.

(Metzios 1994) Metzios, B. G., "Parallel Modeling and Structure of Nonlinear Volterra Discrete Systems," *IEEE Trans. Circuits and Systems,* vol. 41, no. 5, pp. 359–371, May 1994.

(Mulgrew 1994) Mulgrew, Bernard, "Orthonormal Functions for Nonlinear Signal Processing and Adaptive Filtering," *Proc. IEEE ICASSP-94,* vol. 3, Adelaide, South Australia, pp. 509–512.

(Narayan 1983) Narayan, S. Shanker and Peterson, Allen M., "Transform Domain LMS Algorithm," *IEEE Tran. on Acoustics, Speech, and Signal processing,* vol. ASSP-31, pp. 609–615, no. 3, June 1983.

(Newak 1996) Newak, R. D. and Veen, B. D. Van, "Tensor product basis approximation for Volterra filters," *IEEE Trans. Signal Processing,* vol. 44, pp. 36–59, Jan. 1996.

(Nelles 2000) Oliver Nelles, *Nonlinear System Identification,* Springer-Verlag, New York, 2000.

(Ogunfunmi 1994) Ogunfunmi, Tokunbo and Chen, Zhuobin, "Neural Network Algorithms based on the QR Decomposition Method of Least Squares," *Proc. IEEE ICASSP-94,* vol. 3, Adelaide, South Australia, pp. 493–496.

(Ogunfunmi 1998) Ogunfunmi, Tokunbo and Chang, Shue-Lee, "Nonlinear Adaptive Filters: Comparison of the Wiener and Volterra Models," *Proceedings of International Conference on Signal Processing Applications and Technology (ICSPAT),* Toronto, Ontario, Canada, 1998.

(Ogunfunmi 2001) Ogunfunmi, Tokunbo and Chang, Shue-Lee, "Second-order Adaptive Volterra System Identification based on Discrete Nonlinear Wiener Model," *IEE Proceedings-Vision, Image and Signal Processing,* vol. 148, no.1, pp. 21–29, Feb. 2001.

(Ogunfunmi 2006) Ogunfunmi, Tokunbo, Linear Systems (ELEN110) course notes at Santa Clara University, angel.scu.edu, March 2006.

(Ogunnaike 1994) Ogunnaike, Babatunde and Ray, Harmon, W., *Process Dynamics, Modelling and Control,* Oxford University Press, New York, NY, 1994. (See Chapter 12 on Theoretical Process Modeling and Chapter 13 on Process Identification: Empirical Process Modeling).

(Ozden 1996) Ozden, Tahir, Kayran, Ahmet H. and Panyirci, Erdal, "Adaptive Volterra filtering with complete lattice orthogonalization," *IEEE Trans. Signal Processing,* vol. 44, no. 8, pp. 2092–2098, August 1996.

(Papoulis 1991) Papoulis, *Probability, Random Variables, and Stochastic Processing,* Third edition, McGraw-Hill, 1991.

(Paniker 1996) Paniker, T. M. and Mathews, V. J., "Parallel-cascade realization and approximation of truncated Volterra systems," *Proc. IEEE ICASSP,* pp. 1589–1592, 1996.

(Peebles 1987) Peebles, Peyton Z., Jr., *Probability, Random variables, and Random Signal Principles,* McGraw-Hill, 1987.

(Philips 2003) Philips, Parr and Riskin, Signals, Systems and Transforms, Third edition, Prentice-Hall, New York, 2003

(Raz 1998) Raz, Gil M. and Barry Veen, D. Van "Baseband Volterra Filters for Implementing Carrier Based Nonlinearities," *IEEE Tran. Signal Processing,* vol. 46, no. 1, pp. 103–114, Jan. 1998.

(Rice 1966) Rice, J. R., "Experiments on Gram-Schmidt orthogonalization," *Math. of Computation,* vol. 20, pp. 325–328, 1966.

(Rugh WJ 2002) Wilson J. Rugh, *Nonlinear System Theory: The Volterra/ Wiener Approach,* Johns Hopkins University Press, 1981, ISBN 0–8018–2549–0. (Online Web version published 2002.)

(Ryan 1999) Ryan, Francis D., "DSP Implementation of Wiener and Volterra Nonlinear Adaptive Filters for System Identification," *Masters Thesis,* Dept. of Electrical Engineering, Santa Clara University, 1999.

(Sandberg 1983) I. W. Sandberg, "Series expansions for nonlinear systems," *Circuits, Systems and Signal Processing,* vol. 2, no. 1, pp. 77–87, 1983.

(Sayed A 2003) Ali Sayed, *Fundamentals of Adaptive Filters,* IEEE Press, 2003.

(Sayed K 1994) Ali Sayed and Thomas Kailath, *A State Space Approach to Adaptive RLS Filters,* IEEE Signal Processing Magazine, vol. 11, no. 3, pp. 18–60, July 1994.

(Schetzen 1980) Schetzen, M., *Volterra and Wiener Theories of Nonlinear Systems,* New York, Wiley, 1980.

(Schetzen 1981) Schetzen, M., "Nonlinear system modeling based on the Wiener theory," *Proc. IEEE,* vol. 69, no. 12, pp. 1557–1573, Dec. 1981.

(Scott 1997) Scott, Iain and Mulgrew, Bernard, "Nonlinear System Identification and Prediction Using Orthonormal Functions," *IEEE Trans. Signal Processing,* vol. 45, no. 7, pp. 1842–1853, July 1997.

(Silva 1995) Silva, Tomas Oliverira, "On the Determination of the Optimal Pole Position of Laguerre Filters," *IEEE Tran. Signal Processing,* vol. 43, pp. 2079–2087, Sept. 1995.

(Shynk 1989) Shynk, John, "Adaptive IIR Filtering," *IEEE Acoustics, Speech and Signal Processing (ASSP) Magazine,* vol. 6, no. 2, pp. 4–21, April 1989.

(Sjoberg 1995) J. Sjoberg *et al.,* "Nonlinear black box modeling in system identification: A unified overview," *Automatica,* vol. 31, pp. 1691–1724, 1995

(Syed 1994) Syed, M. A. and Mathews, V. J., "Lattice algorithms for recursiveleast squares adaptive second order Volterra filtering," *IEEE Trans.Circuits syst.-II,* vol. 41, Mar. 1994.

(Therrien 1993) Therrien, Charles W. and Hashad, Atalla, "The discrete Wiener model for Representation of Non-Gaussian Stochastic Processes," *Proc. of The 27th Asilomar Conf. Record on Signal, Systems & Computers,* pp. 451–455, Nov. 1993.

(Therrien 1999) Therrien, Charles W., Jenkins, W. Kenneth and Li, Xiaohui, "Optimization of the Performance of Polynomial Adaptive Filters: Making Quadratic Filters Converge like Linear Filters," *IEEE Tran. Signal Processing,* vol. 47, no. 4, April 1999, pp. 1169–1171.

(Therrien 2002) Charles Therrien, "The Lee-Wiener Legacy: A History of the Statistical Theory of Communication," *IEEE Signal Processing Magazine,* pp. 33–44, November 2002.

(Tseng 1993) Tseng, C. H. and Powers, E. J., "Time-domain approach to estimation of frequency-domain Volterra function," *Proc. ATHOS IEEE Workshop System Identification Higher Order Statis.,* Sophia Antipolis, France, Sept. 1993, pp. 6–10.

(Voros 2005) Voros, Jozef, "Identification of Hammerstein Systems with Time-varying Piece-wise Linear Characteristics," *IEEE Trans. Circuits and Systems-II,* vol. 52, no. 12, pp. 865–869, Dec. 2005.

(Walach 1984) Walach, Eugene and Widrow, Bernard, "The Least Mean Fourth (LMF) Adaptive Algorithm and Its Family," *IEEE Tran. Inform. Theory,* vol. IT-30, pp. 275–283, March 1984.

(Westwick K 2003) David T. Westwick and Robert E. Kearney, *Identification of Nonlinear Physiological Systems,* IEEE Press, John Wiley InterScience, 2003.

(Widrow 1985) Widrow, Bernard and Stearns, Samuel D., *Adaptive signal processing,* Prentice-Hall, Englewood, N. J., Second edition, 1985.

(Wiener 1942) Wiener, N., "Response of a nonlinear device to noise," Report No. 129, Radiation Laboratory, M.I.T., April 1942.

(Wiener 1965) Wiener, N., Nonlinear Problem in Random Theory, Wiley and Sons, New York, 1965.

(Zhao 1994) Zhao, Xiao and Marmarelis, Vasilis Z., "Kernel invariance-method for relating continuous-time with discrete-time nonlinear parametric models," *Proc. of IEEE ICASSP,* pp. 533–535, 1998.

INDEX

Signals and Communication Technology

(continued from page ii)

Neuro-Fuzzy and Fuzzy Neural Applications in Telecommunications
P. Stavroulakis (Ed.) ISBN 3-540-40759-6

SDMA for Multipath Wireless Channels
Limiting Characteristics and Stochastic Models
I.P. Kovalyov ISBN 3-540-40225-X

Digital Television
A Practical Guide for Engineers
W. Fischer ISBN 3-540-01155-2

Speech Enhancement
J. Benesty (Ed.)
ISBN 3-540-24039-X

Multimedia Communication Technology
Representation, Transmission and Identification of Multimedia Signals
J.R. Ohm ISBN 3-540-01249-4

Information Measures
Information and its Description in Science and Engineering
C. Arndt ISBN 3-540-40855-X

Processing of SAR Data
Fundamentals, Signal Processing, Interferometry
A. Hein ISBN 3-540-05043-4

Chaos-Based Digital Communication Systems
Operating Principles, Analysis Methods, and Performance Evalutation
F.C.M. Lau and C.K. Tse
ISBN 3-540-00602-8

Adaptive Signal Processing
Application to Real-World Problems
J. Benesty and Y. Huang (Eds.)
ISBN 3-540-00051-8

Multimedia Information Retrieval and Management Technological
Fundamentals and Applications D. Feng, W.C. Siu, and H.J. Zhang (Eds.)
ISBN 3-540-00244-8

Structured Cable Systems
A.B. Semenov, S.K. Strizhakov,and I.R. Suncheley
ISBN 3-540-43000-8

UMTS
The Physical Layer of the Universal Mobile Telecommunications System
A. Springer and R. Weigel
ISBN 3-540-42162-9

Advanced Theory of Signal Detection
Weak Signal Detection in Generalized Obeservations
I. Song, J. Bae, and S.Y. Kim
ISBN 3-540-43064-4

Wireless Internet Access over GSMand UMTS
M. Taferner and E. Bonek
ISBN 3-540-42551-9